BrightRED Revision

Higher PHYSICS

John Taylor

First published in 2008 by:

Bright Red Publishing Ltd
6 Stafford Street
Edinburgh
EH3 7AU

Reprinted (with corrections) in 2010 and 2012.

A CIP record for this book is available from the British Library

ISBN 978-1-906736-02-6

With thanks to Ken Vail Graphic Design, Cambridge (layout) and Ivor Normand (copy-edit)

Cover design by Caleb Rutherford – eidetic

Illustrations by Beehive Illustration (Mark Turner) and Ken Vail Graphic Design

Acknowledgements
Every effort has been made to seek all copyright holders. If any have been overlooked then Bright Red Publishing will be delighted to make the necessary arrangements.

The publishers would like to thank the following for the permission to reproduce the following photographs:
© Peter Menzel/Science Photo Library (p40)
© Sciencephotos/Alamy (p69)

Printed and bound by Charlesworth Press, Wakefield.

CONTENTS

INTRODUCTION

This book provides the content of the Higher Physics syllabus in a **concise** and **attractive** format. Each double page spread presents a new topic which will make learning accessible and digestible. **Key ideas** are illustrated in **colour**. **Key terms** are highlighted to emphasise information that you need to understand to be able to handle the tricky **"explain"** type of examination question. The **graphical** presentations allow **experimental concepts** to be understood through **visual learning**.

By studying this book you should be reinforcing and extending your **knowledge and understanding (KU)** of the concepts of physics. By studying worked examples, practising problems and following the *Let's think about this* features, you will improve your theoretical and **problem solving (PS)** skills. You will be expected to develop **investigative** and **experimental** skills as you progress through the Higher course and this will deepen your understanding of Physics. *Internet links* are also provided throughout the book. These are often not "official" links, but are intended to **illustrate a point**, further **understanding** or provide **additional interest** to the topic.

As you progress through the course, you will find that the **theory** and **concepts** are based on, and illustrated by, **experimental activities**. Physics involves **interaction** between **theory** and **practice** and **pure physics concepts** are illustrated with **applications**.

SYLLABUS AND ASSESSMENT

The book follows the **three mandatory units** of the Higher Physics syllabus:
- Mechanics and Properties of Matter (H)   1 credit
- Electricity and Electronics (H)   1 credit
- Radiation and Matter (H)   1 credit

In addition, a chapter dealing with **scientific data** and **uncertainties in measurement** ensure that **valuable points** are not lost in the **exam** or in the **practical assessment**.

To gain the **overall Higher Physics award** you have to achieve the following stages:
Internal assessment:
- A **PASS** award for each of the **three mandatory units** of the course.
- A **PASS** award for a **report** on **practical** work.
External assessment:
- A **graded** award from an **exam** covering the whole course.

A key to **understanding assessment** is to understand that different assessments have **different types of questions** and questions with **different levels of difficulty**.

Unit assessments

Unit assessments have **shorter questions** and have a higher percentage of **knowledge and understanding** type questions. The unit test will last for about **45 minutes** and have a total of **30 marks**. The pass mark is **60%**. Although the questions are meant to be more straightforward, you will still have to have learned your facts and do simple calculations! The **questions provided** at the **end of each unit** in this book provide a check on your progress over the unit to let you revise the areas you need help with. Try to view the unit assessments as **smaller hurdles** you have to pass before you **prepare** for the **final exam**.

The final exam

The first part of the final exam has 20 multiple choice or objective questions worth 1 mark each. The second part has questions requiring written answers in the form of **short responses** (a few words), **restricted responses** (sentences or a paragraph) or a **numerical calculation**. This part is worth 70 marks so the total paper has 90 marks. This external exam has a time limit of 2 hours 30 minutes.

The final exam has **longer questions** and has a higher percentage of **problem solving** type questions than knowledge & understanding. The exam is **graded** with grade A being the highest award. The exam contains approximately 36 marks awarded for **knowledge & understanding** type questions. The remaining approximately 54 marks cover **problem solving**, **practical skills**, **analytical skills**, **integration** of skills, **application** to **less familiar** or **more complex** contexts and demonstration of **uncertainties** in any unit question. You need to work on more complex problems before your prelim and the final exams.

Practical assessment

There are a range of **practical experiments** to be completed in your Higher Physics course to form your practical report. Your school or college will advise you on suitable experiments and provide you with an *Advice to Candidates* sheet.

If you prepare your report using the following headings you should satisfy the requirements:
- Title
- Aim or Objective
- Apparatus
- Procedure
- Reading/ Results
- Uncertainties
- Conclusion
- Evaluation

All of which, in a nutshell, outlines your Higher Physics course. Good luck!

FORMULAE

Measurement

Analogue: ±0.5 on least division

Digital: ±1 on least significant digit

$$mean = \frac{sum\ of\ values}{number\ of\ values}$$

$$\frac{random}{uncertainty} = \frac{maximum\ value - minimum\ value}{number\ of\ values}$$

$$\%\ uncertainty = \frac{absolute\ uncertainty}{value} \times 100$$

Mechanics and properties of matter

$v = \dfrac{s}{t}$ $v = \dfrac{\Delta s}{\Delta t}$ – see S/Int 2

$v_h = v \cos \theta$ $v_v = v \sin \theta$

$a = \dfrac{\Delta v}{\Delta t}$ $a = \dfrac{v - u}{t}$

$v = u + at$

$s = ut + \frac{1}{2}at^2$

$v^2 = u^2 + 2as$

$s = \dfrac{1}{2}(u + v)t$

Learn newton 1, 2 and 3

$F = ma$ $a = \dfrac{F}{m}$

$F_{un} = U + W$ watch signs

$F = W \sin\theta \pm F_r$ up/down slope

$E_W = Fd$

$P = \dfrac{E}{t}$ $P = Fv$

$E_P = mgh$ $E_k = \frac{1}{2}mv^2$

$Fd = \frac{1}{2}mv^2 + F_{fr}d$

$mg(h_2 - h_1) = F_R d + \frac{1}{2}mv^2$

$p = mv$

$m_1 u_1 + m_2 u_2 = m_1 v_1 + m_2 v_2$

$E_{k\ before} = \frac{1}{2}m_1 u_1^2 + \frac{1}{2}m_2 u_2^2$

$E_{k\ after} = \frac{1}{2}m_1 v_1^2 + \frac{1}{2}m_2 v_2^2$

$Ft = mv - mu$

$\rho = \dfrac{m}{V}$

$P = \dfrac{F}{A}$ $P = \rho gh$

$\dfrac{PV}{T} = constant$

$\dfrac{P_1}{T_1} = \dfrac{P_2}{T_2}$ $\dfrac{V_1}{T_1} = \dfrac{V_2}{T_2}$ $P_1 V_1 = P_2 V_2$

Electricity and electronics

$I = \dfrac{Q}{t}$ $Q = It$ – see S/Int 2

$W = QV$ $V = \dfrac{W}{Q}$

$QV = \frac{1}{2}mv^2$

$\Sigma E = \Sigma IR$

$V = IR$ $R = \dfrac{V}{I}$

$P = \dfrac{E}{t}$ – see Mechanics.

$P = IV = I^2 R = \dfrac{V^2}{R}$

Series: $I = I_1 = I_2 = I_3$
   $V = V_1 + V_2 + V_3$
   $R_T = R_1 + R_2 + R_3$

Parallel: $I = I_1 + I_2 + I_3$
   $V = V_1 = V_2 = V_3$
   $\dfrac{1}{R_T} = \dfrac{1}{R_1} + \dfrac{1}{R_2} + \dfrac{1}{R_3}$

$E = V + v$

$E = V + Ir$ $v = Ir$

$E = IR + Ir$ $V = IR$

$r = \dfrac{v}{I} = \dfrac{E - V}{I}$

$r = -$ gradient

$E = y$ intercept

$V_2 = \dfrac{R_2}{R_1 + R_2} V_S$ – see S/Int 2

$\dfrac{R_1}{R_2} = \dfrac{R_3}{R_4}$ $V \propto \Delta R$

$V_{peak} = \sqrt{2} V_{rms}$ $V_{rms} = \dfrac{V_{peak}}{\sqrt{2}}$

$I_{peak} = \sqrt{2} I_{rms}$ $I_{rms} = \dfrac{I_{peak}}{\sqrt{2}}$

$C = \dfrac{Q}{V}$ $Q = CV$

$E = \dfrac{1}{2}QV = \dfrac{1}{2}CV^2 = \dfrac{1}{2}\dfrac{Q^2}{C}$

C blocks low frequencies & dc

C passes high frequency ac

$\dfrac{V_0}{V_1} = -\dfrac{R_f}{R_1}$ $V_0 = -\dfrac{R_f}{R_1}V_1$

$V_0 = (V_2 - V_1)\dfrac{R_f}{R_1}$

Radiation and matter

$T = \dfrac{1}{f}$ $f = \dfrac{1}{T}$

$v = f\lambda$

path difference $= n\lambda$

path difference $= (n + \frac{1}{2})\lambda$

$n\lambda = d \sin \theta$

$n = \dfrac{\sin \theta_1}{\sin \theta_2}$

$\dfrac{\sin \theta_1}{\sin \theta_2} = \dfrac{v_1}{v_2} = \dfrac{\lambda_1}{\lambda_2}$

$n = \dfrac{v_1}{v_2}$ $n = \dfrac{\lambda_1}{\lambda_2}$

$n_1 \sin \theta_1 = n_n \sin \theta_2$

$n_1 v_1 = n_2 v_2$

$n_1 \lambda_1 = n_2 \lambda_2$

$\sin \theta_c = \dfrac{1}{n}$ $n = \dfrac{1}{\sin \theta_c}$

$I \alpha \dfrac{1}{d^2}$ $I = \dfrac{k}{d^2}$ $I_1 d_1^2 = I_2 d_2^2$

$E = hf$

$I = Nhf$

$I = \dfrac{P}{A}$

$hf = hf_0 + \frac{1}{2}mv^2$

$E_k = hf - hf_0$

$W_2 - W_1 = hf$ $E_2 - E_1 = hf$

$E = mc^2$

$A = \dfrac{N}{t}$

$D = \dfrac{E}{m}$

$H = Dw_R$

$\dot{H} = \dfrac{H}{t}$

SCIENTIFIC DATA

SI UNITS

Système International d'Unités

The units of measurement in the Higher Physics course are based on the **International System** (SI) of units, which consists of SI base units and derived units. The first five SI base units shown here will be used in the Higher Physics course. You will find a full list of the **physical quantities** and their **units** that you learn in Higher Physics on the inside back cover.

Unit	Symbol	Quantity
metre	m	length
kilogram	kg	mass
second	s	time
ampere	A	electrical current
kelvin	K	temperature

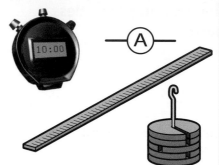

Note:

● Most symbols for units are written in lower case unless named after a person.

● Symbols are not pluralised, nor do they end in a full stop.

● It is good to group numbers in threes with a space between each group and to put a space before the unit: for example, $1\,957\,984\,\text{m}\,\text{s}^{-1}$

 Read more at physics.nist.gov/cuu/Units/units.html

PREFIXES

A prefix produces a multiple of the unit in powers of ten.

Factor		Name	Symbol
10^{12}	1 000 000 000 000	tera	T
10^{9}	1 000 000 000	giga	G
10^{6}	1 000 000	mega	M
10^{3}	1 000	kilo	k
10^{-2}	0.01	centi	c
10^{-3}	0.001	milli	m
10^{-6}	0.000 001	micro	μ
10^{-9}	0.000 000 001	nano	n
10^{-12}	0.000 000 000 001	pico	p

Powers of 10

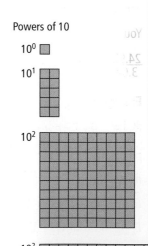

10^{0}
10^{1}
10^{2}
$10^{?}$

Examples

$14\,000\,\text{m} = 14\,\text{km}$

$0.000\,008\,\text{s} = 8\,\mu\text{s}$

$450\,\text{nm} = 0.000\,000\,450\,\text{m}$

$4\,700\,000\,\Omega = 4.7\,\text{M}\Omega$

$15\,\text{THz} = 15\,000\,000\,000\,000\,\text{Hz}$

DON'T FORGET

If you don't watch out for the prefixes in a question, you'll get the wrong answer!

SCIENTIFIC NOTATION

Scientific notation is used by scientists to write large and small numbers.

- The earth's mass is about $5\,973\,600\,000\,000\,000\,000\,000\,000\,$kg
  Easier to write $5.9736 \times 10^{24}\,$kg

- The mass of a proton is $0.000\,000\,000\,000\,000\,000\,000\,000\,001\,673\,$kg
  Easier to write $1.673 \times 10^{-27}\,$kg

On a calculator, e, E or exp often replaces $\times 10^n$. Be careful therefore not to type in $\times 10$ as well!

DON'T FORGET

Calculations: Always check that your answer looks reasonable and possible! Scientific notation is the easiest way for large numbers.

Questions

Multiplication: $(3 \times 10^8) \times (2 \times 10^5) = (3 \times 2) \times 10^{(8+5)}$
$= 6 \times 10^{13}$

Division: $\dfrac{3 \times 10^8}{2 \times 10^5} = \left(\dfrac{3}{2}\right) \times 10^{(8-5)} = 1.5 \times 10^3$

SIGNIFICANT FIGURES

A calculation using measurements **cannot** have more accuracy than the figures of the measurements.

1 m divided by 3 s cannot be $0.333\,333\,3\,$m s^{-1} even if your calculator says so!

As there is only one significant figure in each measurement, the answer should also have only one significant figure.

$$v = \frac{s}{t} = \frac{1}{3} = 0.3\,\text{m s}^{-1}$$

3 s, 3.0 s, 3.00 s, and 3.000 s all have the same value, but 3.000 s we can assume has been measured more precisely. 1.00 m is more precise than 1 m.

$$v = \frac{s}{t} = \frac{1.00}{3.000} = 0.333\,\text{m s}^{-1}$$

If your calculator produces too many significant figures, you can **carry these through** until the end of your working, but you must then **remember to round up**.

You should **round up to the smallest number of significant figures** in the measurements.

DON'T FORGET

Just round up at the end of a calculation.

$\dfrac{24.92}{3.64} =$ ⬛ $6.846\,153\,8$ ⬛   Rounded to 3 significant figures   6.85

Examples

30 has 1 significant figure.

0.007 020 0 has 5 significant figures.

5400 has 2 significant figures.

30.0 has 3 significant figures.

5.40×10^3 has 3 significant figures.

LET'S THINK ABOUT THIS

1 There is usually a 0.5 mark for the unit after a numeric answer. All these 0.5 marks add up!

2 Missing a prefix in a question puts your answer out by powers of 10!

3 You can do a rough check on answers just using the powers of 10 in scientific notation.

4 Keep an additional significant figure in intermediate values to calculations.

5 Zeros are important!

UNCERTAINTY IN MEASUREMENT

When a **measurement** is made of a physical quantity, it will **always** be liable to an **uncertainty**.

READING UNCERTAINTIES

Scale-reading uncertainties indicate the accuracy to which an instrument scale can be read. Scale-reading uncertainties exist in both analogue and digital scales.

Analogue Scales

Reading uncertainty for an **analogue scale** is: ±0.5 of the least division

This allows us to read to the **nearest half division**.

Example: How well can you read these scales?

The smallest division is 1 V, half this = 0.5 V
The reading is: **3 ± 0.5 V**

The smallest division is 0.1 A, half this = 0.05 A
The reading is: **0.35 ± 0.05 A**

An exception is made for **wide divisions** where a more reasonable estimate would be: ±0.2 of the least division

This allows us to read to the nearest one fifth of a wide division.

Example: How well can you read this scale?

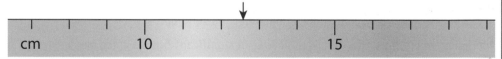

The smallest division is 1 cm, one fifth of this = 0.2 cm
The reading is: **12.6 ± 0.2 cm**

Digital Scales

Reading uncertainty for a **digital scale** is: ±1 on the least significant digit

This allows us to read the **least significant digit**.

Example: What is the reading uncertainty in this digital stop-watch?

DON'T FORGET

How well a scale can be read is only one source of uncertainty.

The least significant digit is 0.01 s.
The reading is: **3.26 ± 0.01 s**

RANDOM UNCERTAINTIES

When a measurement of a physical quantity is repeated, we usually see different readings. There is **equal probability** that each measurement made **is higher or lower** than the 'true' value. **Repeated measurements** of a physical value are **desirable**. The mean of repeated measurements is the **best estimate of a 'true' value** being measured.

The mean is found from the sum of the measurements divided by the **number** of measurements.

$$\text{mean} = \frac{\text{sum of values}}{\text{number of values}}$$

Example falling ball; $t = 1.48, 1.32, 1.38, 1.37, 1.40\,\text{s}$

$$\text{mean} = \frac{6.95}{5} = 1.39\,\text{s}$$

Repeating measurements reduces uncertainty. The greater the number of repeats, the smaller the uncertainty.

Uncertainty can be estimated by dividing the **range** by the **number** of measurements.

$$\text{random uncertainty} = \frac{\text{maximum value} - \text{minimum value}}{\text{number of values}}$$

Example falling ball; uncertainty $= \frac{1.48 - 1.32}{5} = \frac{0.16}{5} = 0.032 = 0.03$

Experimental measurements should be expressed in the form:

value ± uncertainty

Example falling ball; $t = 1.39 \pm 0.03\,\text{s}$

SYSTEMATIC UNCERTAINTIES

A **systematic uncertainty** is an error which affects all the measurements in the **same way**. A systematic uncertainty will make the readings either all **too high** or all **too low**.

Where a systematic effect is present, the **mean value** of the measurements will be **offset** from the 'true' value of the physical quantity being measured.

Faults in the apparatus:

A meter may not be set to its zero value properly.

A ruler may be used without allowing for the end space.

Experimental technique:

A scale may always be read from one side rather than directly in front.

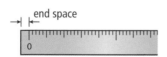

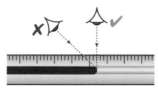

LET'S THINK ABOUT THIS

1 On an analogue scale with very small divisions, we may only be able to read to 1 division.

2 To increase the accuracy of a measurement, take more repeats.

3 On a graph: a straight line which nearly goes through the origin is likely to have had measurements offset in one way.

4 What is the difference between random uncertainties and systematic effects?

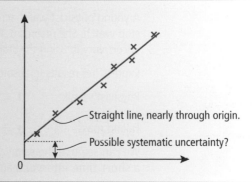

Straight line, nearly through origin.

Possible systematic uncertainty?

DEALING WITH UNCERTAINTIES

The **measurement** of any physical quantity is liable to **uncertainty**.

The experimenter has to **estimate** the **true value** (e.g. by calculating the mean) and then **estimate** the **uncertainty in the mean** (e.g. by using the random uncertainty equation).

ABSOLUTE AND PERCENTAGE UNCERTAINTIES

Absolute Form

The uncertainty can be expressed with the same units as the value measured. This is known as the **absolute uncertainty**. The results are written as shown:

estimated value $\longleftarrow$
$$2.78 \pm 0.03 \, \text{mA}$$
$$5.6 \pm 0.1 \, \mu\text{V}$$
$$15.23 \pm 0.50 \, \text{s}$$
$\longrightarrow$ absolute uncertainty and unit

Are you certain?

It's the truth!

Percentage Form

Uncertainty can also be expressed in percentage form. Percentage uncertainty allows us to indicate how precise a value is. The **percentage uncertainty** has to be calculated:

$$\% \text{ uncertainty} = \frac{\text{absolute uncertainty}}{\text{measurement}} \times 100$$

Check you can obtain these percentage uncertainties for the previous absolute uncertainties:

2.78 mA, 1%   5.6 μV, 2%   15.23 s, 3%

Example

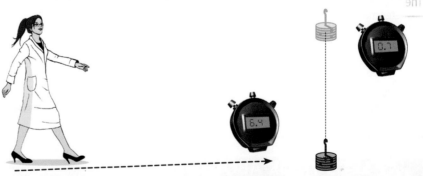

A young physicist walks across the room. She has measured her time to do this with a stop-watch. She records 6.4 s. She then times a falling mass to take 0.7 s. She estimates her uncertainty in using the stop-watch to be 0.2 s.

Express her results in absolute form and calculate the percentage uncertainty for each.

Physicist's walk: $t = 6.4 \pm 0.2$ s   % uncertainty $= \frac{0.2}{6.4} \times 100 = 3\%$

Falling mass: $t = 0.7 \pm 0.2$ s   % uncertainty $= \frac{0.2}{0.7} \times 100 = 29\%$

We can see that the same absolute uncertainty has a **greater percentage effect** when timing a **short time interval** than when timing over a long time.

COMBINING UNCERTAINTIES

In doing an experiment, several variables may be measured then the final value calculated using an equation.

The final numerical result of an experiment should be given as: final value ± uncertainty

1 The **final value** should be calculated from the measurements using the required equation in the normal way.

The final or **overall uncertainty** needs to be calculated.

2 Calculate the percentage uncertainty for **each** measurement.
3 The **largest** percentage uncertainty is the most significant.
4 The **largest percentage uncertainty is a good estimate of the percentage uncertainty in the final numerical result** of the experiment.
5 **Use** this percentage uncertainty to calculate the **absolute uncertainty** from the **final value**.

Example

A physicist is asked to find out how much energy is used when a kettle is boiled. He takes the following measurements:

Voltage of supply $V = 225 \pm 25\,V$
Current drawn $I = 9.5 \pm 0.5\,A$
Time taken $t = 125 \pm 5\,s$

Energy $E = ItV = 9.5 \times 125 \times 225 = 267\,188\,J$

Percentage uncertainties:

V $\frac{25}{225} \times 100 = 11\%$

I $\frac{0.5}{9.5} \times 100 = 5\%$

t $\frac{5}{125} \times 100 = 4\%$

The largest percentage uncertainty is **11%**.

The final uncertainty: 11% of 267 188 J = 29 391 J

Remember to round to the fewest significant figures!

The final result: Energy,   **E = 0.27 ± 0.03 MJ**

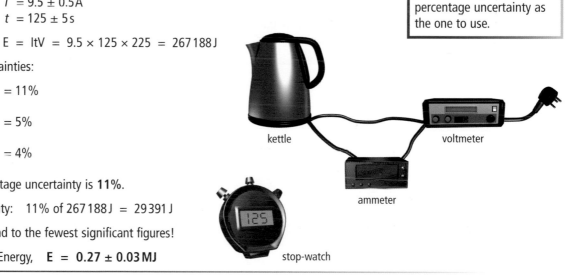

kettle

voltmeter

ammeter

stop-watch

> **DON'T FORGET**
>
> Select the largest percentage uncertainty as the one to use.

LET'S THINK ABOUT THIS

1 You are asked to time the period of a pendulum. How would you reduce the uncertainty in the measurement of the period (time for one swing)?

Do not try to time one swing. The time will be very short. The percentage uncertainty will be high. Time a number of periods (say 10) and divide the result by the number of swings. The absolute uncertainty is then also divided when you calculate one period.

2 Whenever you are asked to do a formal report of an **investigation**, you should think of all the uncertainties in each measurement and remember to combine them if you calculate a value.

SCALARS, VECTORS AND DISPLACEMENT

SCALAR AND VECTOR QUANTITIES

In physics, we measure, calculate and study relationships between **physical quantities**.

The two types of physical quantity are **scalar** and **vector**.

Scalars

A **scalar quantity** is defined by its **magnitude** alone. This means that it only has size. Along with the magnitude, its **unit** will be given. Scalar quantities are added using basic arithmetic.

Vectors

A **vector quantity** is defined by both its **magnitude** and its **direction**.

To add two vectors together, we must consider the effect of the direction as well as the magnitude. To add vectors together, we have to use scale diagrams or some mathematics.

Scalars and vectors are quantities that you will use in mechanics:

Scalars	Vectors
distance	displacement
speed	velocity
time	acceleration
mass	force
energy	momentum
power	impulse

DON'T FORGET

If we ignore the direction of a vector, the result will be wrong!

COMBINING VECTORS

Scale diagram method

A vector can be a line drawn to scale with an arrow to show the direction.

To combine vectors:

1 Choose and write down a **scale**.
2 Add vectors drawn '**head to tail**'.
3 Draw the **resultant** from '**start to finish**'.
4 Measure both the **magnitude** and the **direction**.

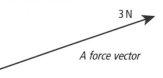
3 N

A force vector

LET'S THINK ABOUT THIS

Mathematical methods (additional only if you are familiar with these equations!).

at 90°: Pythagoras: $c = \sqrt{a^2 + b^2}$ and trigonometry: $\tan \theta = \dfrac{\text{Opposite}}{\text{adjacent}}$

acute or obtuse: Sine rule: $\dfrac{a}{\sin A} = \dfrac{b}{\sin B} = \dfrac{c}{\sin C}$

Cosine rule: $a^2 = b^2 + c^2 - 2bc \cos A$

DISTANCE AND DISPLACEMENT

Distance is a **scalar** quantity. The symbol for distance is **d**.
Distance is defined by a number and its unit, the metre, m.
   e.g. 'The length of the school laboratory is **8 m**' refers to a distance.

Displacement is a **vector** quantity. The symbol for displacement is **s**.
Displacement is the distance travelled in a stated direction from the starting point.
   e.g. 'An explorer walks **20 km due North** from base camp' is referring to a displacement.

Examples

1 A pupil travels East 10 m along the school corridor but is sent back 7 m for running!

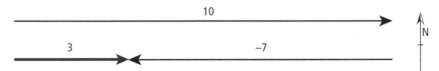

distance travelled, d = 10 + 7 = **17 m**

displacement is 10 + (−7) = **3 m East**.

If this is repeated, the pupil's displacement will still only be **6 m East**.

2 A car drives 4 km North then drives 3 km West. What is the resultant displacement?

Using Pythagoras and trig:

$x^2 = 4^2 + 3^2$

$x = 5\,km$

$\tan \theta = \dfrac{3}{4}$ $\theta = 37°$

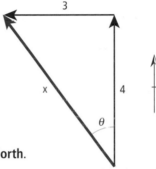

displacement, **s = 5 km @ 37° to the West of North**.

Use a ruler and protractor to check this by scale measurement.

3 A runner goes round the race track three times. If the track has a length of 220 m, what is the runner's distance travelled and displacement?

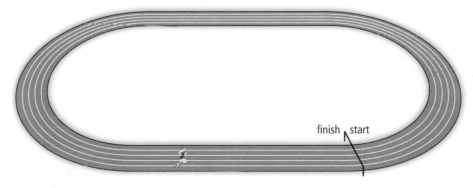

distance travelled
d = 3 × 220 = 660 m.

displacement
s = 0 m.

Displacement is different from distance.

DON'T FORGET

Displacement is the distance from start to finish in a direction.

VELOCITY, ACCELERATION AND FORCE

SPEED AND VELOCITY

Speed is a **scalar** quantity. Speed is the distance travelled in unit time.

$$\text{speed} = \frac{\text{distance}}{\text{time}}$$

Speed only has **magnitude**, no direction.

e.g. 'The car was going faster than **30 m s^{-1}'** refers to a speed.

Velocity is a **vector** quantity. Velocity is the displacement per unit time.

$$\text{velocity} = \frac{\text{displacement}}{\text{time}} \quad \text{or} \quad v = \frac{s}{t}$$

Velocity has **magnitude** and direction.

e.g. 'The car was travelling at **30 m s^{-1} from Glasgow to Edinburgh'** refers to a velocity.

Example

A sailor sets his boat on a heading of North at 5 m s^{-1} through the sea. The tide is moving at 2 m s^{-1} in a South-East direction.

Find the boat's resultant velocity over the ground.

Scale: 1 cm ≡ 1 m s^{-1}

DON'T FORGET

If a bearing has been used for the direction, 000° is North.

DON'T FORGET

... the direction, or you could lose half the marks!

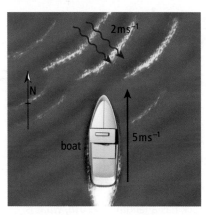

Head-to-tail diagram:

Resultant from start to finish

ruler

protractor

Measure resultant, x = 3.9 cm

Measure θ = 22°

Resultant velocity, **v = 3.9 m s^{-1} @ 022°**

RESOLUTION OF VELOCITIES

A vector quantity can be **resolved** into two **components** at right angles.

e.g. An object leaves the ground with velocity, v, at angle θ.

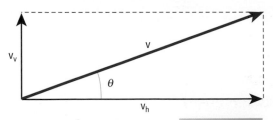

Horizontal component: $v_h = v \cos \theta$

Initial vertical component: $v_v = v \sin \theta$

ACCELERATION AND FORCE

Acceleration and **force** are both **vector** quantities. Acceleration is directly proportional to force.

The **downward force** of gravity produces **acceleration downwards**.

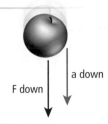

a down

F down

The acceleration of an object will be in the **same direction** as the resultant unbalanced force.

RESOLUTION OF FORCES

As **force is a vector**, it can be replaced by two forces which are at **right angles** to each other.

A roller is pulled across the grass with a force, F, at an angle, θ.

The two components of this pull are:

the vertical component: $\quad F_v = F \sin \theta$

the horizontal component: $\quad F_h = F \cos \theta$

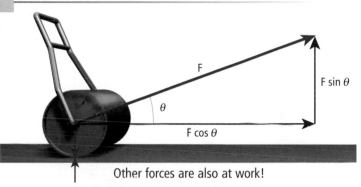

F

F sin θ

θ

F cos θ

Other forces are also at work!

ADDING FORCES

The **resultant** of a number of forces is the single force, which has the **same effect** as the sum of the individual forces.

Force is a vector quantity, and the resultant force must be found by **vector addition** (scale diagram or by calculation).

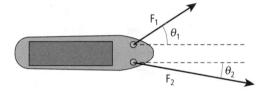

F_1

θ_1

F_2

θ_2

Scale diagram:

Measure the magnitude and direction.

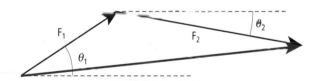

F_1

θ_1

F_2

θ_2

> **DON'T FORGET**
>
> It can be easier to use the scale diagram method!

By calculation:

Resolve each force into two components.

Add the horizontal and vertical separately.

Horizontal components: $F_1 \cos \theta_1 + F_2 \cos \theta_2$

Vertical components: $F_1 \sin \theta_1 + F_2 \sin \theta_2$

Now use Pythagoras and trig to calculate resultant.

$$a^2 = b^2 + c^2 \qquad \tan \theta = \frac{opp}{adj}$$

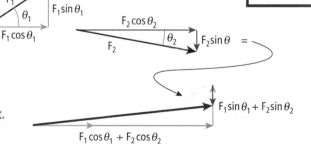

F_1

θ_1

$F_1 \sin \theta_1$

$F_1 \cos \theta_1$

$F_2 \cos \theta_2$

F_2

θ_2

$F_2 \sin \theta$ =

$F_1 \sin \theta_1 + F_2 \sin \theta_2$

$F_1 \cos \theta_1 + F_2 \cos \theta_2$

LET'S THINK ABOUT THIS

1 The cosine component is adjacent to the angle.

2 If you have the 'wrong' angle x, you can use $\theta = (90 - x)$.

3 When adding more than two forces, or if the forces are not at right angles, it is best to use the scale diagram method.

VELOCITY AND ACCELERATION EQUATIONS

MEASURING VELOCITY

A good **estimate** of the **magnitude** of an **instantaneous velocity** can be obtained by measuring a small change in **displacement** over a small change in **time**.

$$v = \frac{\Delta s}{\Delta t}$$

$(\Delta = \text{change in})$

- Using a ruler, **measure** the **length** of card.

- The light gate and motion computer measure the short **time** interval for the card to cut the light beam.

- The velocity equation can then be used to **calculate** the magnitude of the velocity.

MEASURING ACCELERATION

Acceleration is **the change in velocity per unit time**.

1 A good **estimate** of the **magnitude** of **acceleration** can be obtained by measuring a small change in **velocity** over a small change in **time**.

$$a = \frac{\Delta v}{\Delta t} \quad \text{or} \quad a = \frac{v - u}{t}$$

- The light gate and motion computer measures the **times** for card 1, card 2 and the time interval.

- u and v are calculated using the **velocity** equation.

- The acceleration equation can then be used to **calculate** the magnitude of the **acceleration**.

2 A similar method uses **two light gates** and **one card length** to calculate the average acceleration. In this method, you may need to measure the time interval between velocities *u* and *v* with a **stop-watch**.

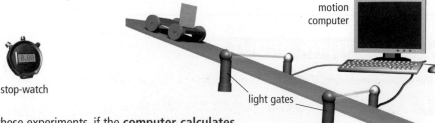

In these experiments, if the **computer calculates** the **acceleration**, you will need to **enter** the **length of card**.

EQUATIONS OF MOTION

For an object which has **constant acceleration** in a **straight line**, we can derive alternative equations or relationships which include displacement, s, as well as u, v, t and a.

1 **Rearrange the acceleration equation:**

$$a = \frac{v - u}{t} \quad \Rightarrow \quad \boxed{v = u + at} \quad \ldots\ldots \quad \text{equation 1}$$

2 **Equation derived from a *v/t* graph:**

An object is accelerating from initial velocity *u* to a final velocity *v* in time *t*.

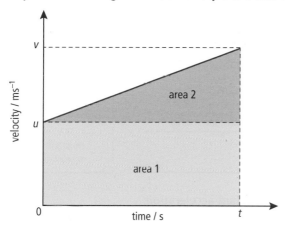

www.diracdelta.co.uk has an online *equations of motion* calculator. Just visit the site and enter *equations of motion* into its search box.

Displacement, s, = area under a *v/t* graph

$s = $ area 1 + area 2
$s = (ut) + (\frac{1}{2}(v - u)t)$
$s = ut + \frac{1}{2}(at)t$ (as equation 1 gives $v - u = at$)
$\boxed{s = ut + \frac{1}{2}at^2}$ $\ldots\ldots$ equation 2

3 **Equation combined from equations 1 and 2:**

From equation 1 $v = u + at$
square both sides $v^2 = (u + at)^2$
    $v^2 = u^2 + 2uat + a^2t^2$
    $v^2 = u^2 + 2a(ut + \frac{1}{2}at^2)$ (tricky step: $1 = 2 \times \frac{1}{2}$!)
From equation 2 $(ut + \frac{1}{2}at^2) = s$
    $\boxed{v^2 = u^2 + 2as}$ $\ldots\ldots$ equation 3

DON'T FORGET

You are expected to be able to derive these important equations.

LET'S THINK ABOUT THIS

1 Equation checkpoint:

Eqn	a	u	v	s	t
1	✓	✓	✓	0	✓
2	✓	✓	0	✓	✓
3	✓	✓	✓	✓	0

Use the equation that fits best.

(List variables then choose equation.)

2 Vectors: *a, u, v* and *s* are vectors. Not *t*.

3 Directions: Take original direction / forwards as positive (+), backwards as negative (–). Take up as positive (+), down as negative (–).
Note: If you have been taught the opposite to this, then it's OK, but stick to one method.

4 Gravity pulls down whether an object is going up or down, so $a = g = -9.8\,\text{m s}^{-2}$ always.

5 Rocket: acceleration up is positive due to thrust up.

GRAPHS AND CALCULATIONS

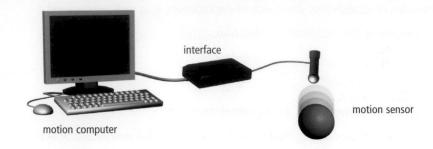

interface

motion sensor

motion computer

GRAPHS OF MOTION

1 These three graphs show the same **constant velocity**:

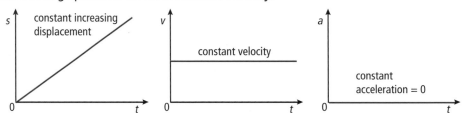

2 These three graphs show the same **constant acceleration**:

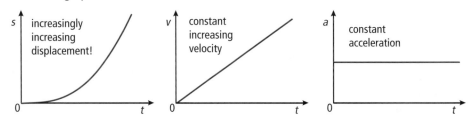

3 Object accelerates from rest, maintains constant velocity, then decelerates to rest:

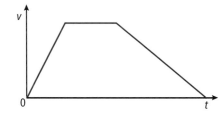

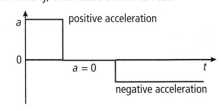

4 Ball thrown upwards until it returns:

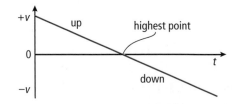

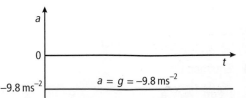

5 Bouncing ball, ball dropped from hand:

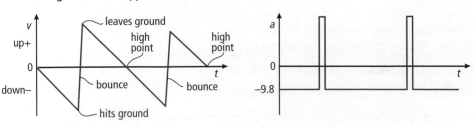

CALCULATING MOTION

1 A vehicle runs down a track, from rest, a distance of 0.5 m in a time of 0.6 s. Calculate the vehicle's **acceleration**.

$s = 0.5\,m$
$t = 0.6\,s$
$u = 0\,ms^{-1}$
$s = ut + \frac{1}{2}at^2$
$0.5 = 0 + \frac{1}{2}a\,0.6^2$
$a = 2.8\,ms^{-2}$

acceleration, $a = 2.8\,ms^{-2}$

0.5 m

2 A vehicle runs down a track, from rest, a distance of 0.8 m before a computer timer measures its velocity to be 1.55 ms⁻¹. Calculate the vehicle's **acceleration**.

$u = 0\,ms^{-1}$
$s = 0.8\,m$
$v = 1.55\,ms^{-1}$
$v^2 = u^2 + 2as$
$1.55^2 = 0^2 + 2a\,0.8$
$a = 1.5\,ms^{-2}$

acceleration, $a = 1.5\,ms^{-2}$

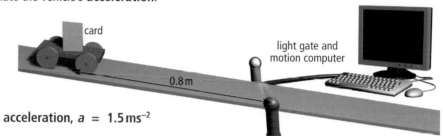

card

light gate and motion computer

0.8 m

3 A plane, taxiing down the runway at 20 ms⁻¹, then accelerates at 3 ms⁻² for 20 s before take-off. What was the plane's take-off **velocity**, and what was its **distance** travelled while accelerating?

20 ms⁻¹   30 s

$u = 20\,ms^{-1}$
$a = 3\,ms^{-2}$
$t = 20\,s$

$v = u + at$ $s = ut + \frac{1}{2}at^2$
$v = 20 + 3 \times 20$ $s = (20 \times 20) + (\frac{1}{2} \times 3 \times 20^2)$
$\mathbf{v = 80\,ms^{-1}}$ $\mathbf{s = 1000\,m}$

4 A ball is thrown upwards with an initial velocity of 6 ms⁻¹. Find the **height** reached and **how long** this takes.

$u = 6\,ms^{-1}$
$v = 0\,ms^{-1}$
$a = -9.8\,ms^{-2}$
$s = ?$
$t = ?$

$v = u + at$ $v^2 = u^2 + 2as$
$0 = 6 + (-9.8 \times t)$ $0^2 = 6^2 + 2 \times (-9.8) \times s$
$t = 0.61\,s$ $s = 1.8\,m$

time to top, $t = 0.61\,s$ **height, $h = 1.8\,m$**

6 ms⁻¹

DON'T FORGET

Practise many more problems with these equations!

LET'S THINK ABOUT THIS

Area under a *velocity/time* graph = **displacement**

Area under an *acceleration/time* graph = **velocity**

Gradient of a *velocity/time* graph = **acceleration**

Equations of motion apply only to motion in a **straight line** (e.g. not a pendulum).

PROJECTILES

Previous problems have investigated objects thrown up, objects dropped vertically and objects going straight on a slope. Another type of motion results when an object is projected **horizontally** or at an **angle**. This type of motion is known as **projectile motion**.

HORIZONTAL PROJECTION

Compare an **object dropped vertically** with an **object projected horizontally** (which should both take the **same time** to hit the ground):

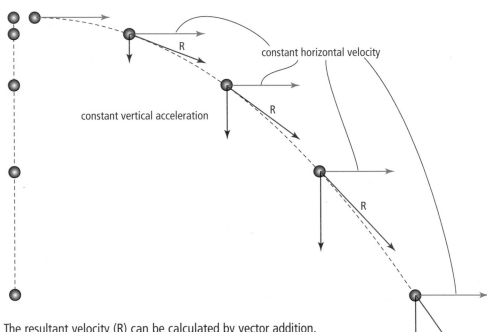

constant horizontal velocity

constant vertical acceleration

> **DON'T FORGET**
>
> Give the magnitude and direction!

The resultant velocity (R) can be calculated by vector addition.

The motion of a projectile is made up of **two** independent **components**:

1 Constant horizontal velocity. Use the equation $\quad v = \dfrac{s}{t}$

2 Constant vertical acceleration. Use the equations of motion:

$$v = u + at \qquad s = ut + \tfrac{1}{2}at^2 \qquad v^2 = u^2 + 2as$$

> **DON'T FORGET**
>
> Always take care with directions!

When solving problems, list the components **independently** and clearly. **Note** that the **time** of flight is the **same** both vertically and horizontally. **Note** that the initial vertical component may be $u = 0\,\text{ms}^{-1}$. You should know $a = g = -9.8\,\text{ms}^{-2}$.

Example

An object is projected horizontally off a cliff with a velocity of $5\,\text{ms}^{-1}$. If the cliff is $6\,\text{m}$ high, calculate how far out the object lands from the base of the cliff.

Vertically

$s = -6\,\text{m}$
$u = 0\,\text{ms}^{-1}$
$a = g = -9.8\,\text{ms}^{-2}$
$t = ?$

$s = ut + \tfrac{1}{2}at^2$
$-6 = 0 + \tfrac{1}{2} \times (-9.8) \times t^2$
$t = 1.11\,\text{s}$

Horizontally

$s = ?$
$v = 5\,\text{ms}^{-1}$
$t = ?$

Then:
$s = vt = 5 \times 1.11 = 5.55\,\text{m}$

Distance from base is 5.55 m

Search *projectiles* at www.physics.org for more information.

PROJECTION AT AN ANGLE

An object may be **projected** into the air at an angle. For example, an arrow is fired at an **angle from the horizontal**.

The initial velocity of the projectile must be resolved into horizontal and vertical **components**.

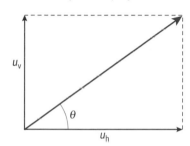

$$u_h = u \cos \theta \qquad u_v = u \sin \theta$$

The **horizontal component** of **velocity** is **constant**.

There is **constant vertical acceleration** due to the force of gravity acting downwards.

Assuming **air resistance** is **negligible**, the trajectory will be **symmetrical**:

- The final vertical component of velocity = – the initial vertical component of velocity.
$$v_v = -u_v$$
- The time of flight = 2 × time to the highest point.
- Vertical component of velocity at the top = $0\,ms^{-1}$ (but still has horizontal component).
- Vertically, $a = g = -9.8\,ms^{-2}$.

Example

A golfer strikes a ball which moves off at an angle of 25° to the ground at $60\,ms^{-1}$. The ball lands 6 s later. What distance does the ball travel and what is the vertical component of its velocity just before it hits the ground?

Horizontally: $u_h = u \cos \theta = 60 \times \cos 25° = 54.4\,ms^{-1}$
$s = vt = 54.4 \times 6 = 326.3\,m$ **distance = 326.3 m**

Vertically: $u_v = u \sin \theta = 60 \times \sin 25° = 25.4\,ms^{-1}$ upwards.
Final vertical component of velocity = **25.4 ms⁻¹ downwards**

LET'S THINK ABOUT THIS

The **horizontal** and the **vertical components** are completely **independent**. That is, one does not affect the other. Nothing, except air resistance, changes the horizontal component, but gravity changes vertical motion. The time is, however, the same for both components.

NEWTON'S LAWS

BALANCED AND UNBALANCED FORCES

DON'T FORGET

Force is a vector.

Balanced forces:

Forces which are **equal in size** but **opposite in direction** are called **balanced forces**.

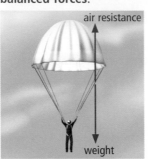

air resistance

weight

Unbalanced forces:

The **resultant of forces** which do not cancel out is known as the **unbalanced force**.

An **unbalanced force** causes **acceleration**.

The dragster accelerates because the engine thrust is greater than the resistive forces.

thrust   drag

NEWTON'S FIRST LAW OF MOTION

N1: An object will **remain at rest** or will **remain at constant velocity** unless acted on by an unbalanced force.

DON'T FORGET

Balanced forces and no force gives the same effect.

Newton's First Law is a **no-force** law.

In space travel, no force is needed to keep moving. If no (e.g. resistive) forces act on an object, it continues at a steady speed in a straight line.

The space probe just keeps on going.

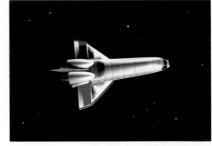

Friction

To keep a vehicle moving, we normally have to keep applying a force. Why is this?

An object in motion experiences **resistive forces** that **increase with velocity**. These are known as the forces of friction. **Friction** always acts **against** the **direction** of the **motion**.

engine thrust   friction

DON'T FORGET

The force of **friction** increases with **velocity**.

Where the **applied force** and **friction** are **balanced**, then **Newton's First Law** tells us that the object will remain at **constant velocity**.

 Mini quiz on Newton's laws: www.quia.com/jq/19675.html

NEWTON'S SECOND LAW OF MOTION

N2: The **acceleration** of an object **varies directly** with the **unbalanced force** and **inversely** with its **mass**.

Thus: $a \propto \dfrac{F}{m}$ $a = k\dfrac{F}{m}$

The Newton defined

The unit of force is the **newton**. **1 N** is defined as the resultant force, which causes a mass of **1 kg** to accelerate at **1 ms^{-2}**.

Substituting into the above equation: $1 = k\dfrac{1}{1}$ $k = 1$

$a = \dfrac{F}{m}$ or $\boxed{F = ma}$ known as Newton's Second Law equation (**N2**).

Newton's Second Law tells us that **force causes acceleration**.

Example

A 100 kg vehicle accelerates. The small engine force is 200 N, but friction exerts 50 N. Find the acceleration.

$a = \dfrac{F_{un}}{m} = \dfrac{200 - 50}{100} = 1.5\,\text{ms}^{-2}.$

DON'T FORGET

The **unbalanced** force causes acceleration, so

$F_{un} = ma$

NEWTON'S THIRD LAW OF MOTION

N3: If A exerts a force on B, then B exerts an **equal but opposite** force on A.

Newton's Third Law tells us that forces exist in **pairs**. These **Newton pairs** are **equal in size** but **opposite in direction**.

The **rocket** pushes the **fuel** away.

The **fuel** pushes the **rocket** away.

Note that both forces occur at the **same time**.

force of fuel on rocket

force of rocket on fuel

LET'S THINK ABOUT THIS

1   The previous spreads dealt with the **measurement of motion**, or **KINEMATICS**.

2   These next spreads deal with the **forces causing motion**, or **DYNAMICS**.

3   Constant velocity = constant speed in a straight line. If direction changes, velocity changes.

4   Acceleration is F_{un}!

5   List some other examples of Newton pairs.

FREE BODY DIAGRAMS

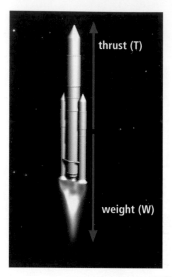

thrust (T)

weight (W)

When **several forces** act on a body, a **free body diagram** is used to show these forces.

ROCKETS

Adding the forces together:

$$F = T + W$$

If **Thrust** is **positive**, **Weight** will be **negative**.

Mass decreases as the fuel is rapidly used implies $\Big\}$ $W = mg$

Gravitational field strength decreases with height implies $\Big\}$ Weight decreases.

The air resistance has been considered as negligible but would be negative (zero at take-off!)

Example

A rocket of mass 9000 kg at take-off has a thrust of 180 000 N. Calculate the acceleration.

$$F = T + W = 180\,000 + (9000 \times -9.8) = 91\,800\,N$$

$$a = \frac{F_{un}}{m} = \frac{91\,800}{9000} = 10.2\,ms^{-2}$$

PERSON IN A LIFT

upward force (U)

weight (W)

Upward force (U): Let upward force be a positive number

Unbalanced force $F_{un} = U + W$

Weight (W): Let weight down be a negative number

Acceleration $a = \dfrac{F_{un}}{m}$

If you stand on bathroom scales in a lift, your weight appears to change when the lift accelerates. Your weight doesn't change; the scales read your weight plus the unbalanced force, causing acceleration (this gives the upward force).

Examples

1   If the lift is **stationary** or moving at a **constant speed**:

$$F_{un} = U + W = 0\,N$$

If mass = 70 kg, $W = mg = 70 \times -9.8 = -686\,N$, so $U = +686\,N$.

$$F_{un} = U + W = 686 + (-686) = 0\,N, \Rightarrow \text{scales read } 686\,N.$$

2   If the lift is **accelerating up**:

Upward force increases: upward force > weight, so unbalanced force and acceleration are **positive**.

If the lift accelerates up at 2 ms^{-2}, accelerating force $F_a = ma = 70 \times 2 = 140\,N$.

Total upward force = 686 + 140 = 826 N, $\Rightarrow$ scales read 826 N.

3   If the lift is **accelerating down**:

Upward force decreases: upward force < weight, so unbalanced force and acceleration are **negative**.

If the lift accelerates down at 2 ms^{-2}, the scales will read 546 N. Check this yourself!

TRAINS, CARAVANS AND PARCELS

This theory applies to **connected objects** that are being **pushed** or **pulled**.

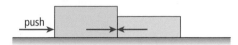

Example

T = Tension. Take friction, F_R, for each wagon to be 100 N.

$$a = \frac{F}{m} = \frac{P - F_R}{m} = \frac{1300 - 300}{100} = 10\,\text{ms}^{-2}$$

$$F_B = ma = 80 \times 10 = 800\,\text{N} \quad \Rightarrow \quad T_B = F_B + F_R = 800 + 200 = 1000\,\text{N}$$

$$F_C = ma = 20 \times 10 = 200\,\text{N} \quad \Rightarrow \quad T_C = F_C + F_R = 200 + 100 = 300\,\text{N}$$

INCLINED PLANES

The **component** of weight **into** (perpendicular to) a slope is balanced by the **normal reaction**.

The **component** of weight **down** (parallel to) a slope causes **acceleration**.

Consider these **frictionless** inclined planes:

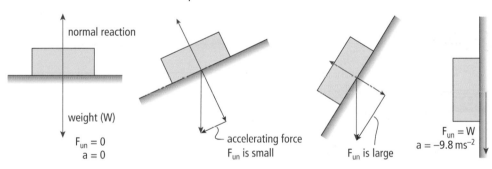

Calculating components of weight:

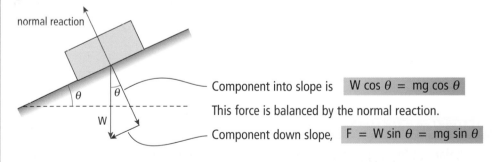

Component into slope is $\boxed{W \cos \theta = mg \cos \theta}$

This force is balanced by the normal reaction.

Component down slope, $\boxed{F = W \sin \theta = mg \sin \theta}$

> **DON'T FORGET**
>
> Weight down is split into two components at right angles.

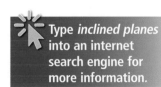

Type *inclined planes* into an internet search engine for more information.

LET'S THINK ABOUT THIS

1 In a rope, **tension** is the equal and opposite force to the pulling force.

2 Inclined plane with friction, object moving **up** (forces in **same** direction): $F = W \sin \theta + F_R$

3 Inclined plane with friction, object moving **down** (forces **opposite**): $F = W \sin \theta - F_R$

4 Friction is always **against** the direction of motion.

WORK, ENERGY AND POWER

WORK, ENERGY AND POWER

Work is done only when **energy is transferred**.

Work done = Energy transferred

$W = E_W$ (Note that the symbols W and E_W may be interchanged in an equation.)

Work is done on an object when a **force** is used to move the object a certain **distance**.

Work is a **scalar** quantity. $E_W = Fd$

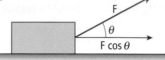

$E_W = F \cos\theta \times d$

Power is the **rate** of **doing work**.

Power is the **work done** in **unit time**. $P = \dfrac{E}{t}$

$P = \dfrac{E_w}{t} = \dfrac{Fd}{t} = Fv$ $P = Fv$

Example

A parcel is pulled along a distance of 5 m at a speed of $4\,ms^{-1}$ by a force of 50 N.

$E_W = Fd = 50 \times 5 = 250\,J$
$P = Fv = 50 \times 4 = 200W$.

POTENTIAL ENERGY

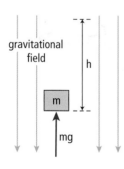

A mass is lifted against the gravitational field: $E_W = Fd = mg \times h$

Lifting force and weight are balanced: $E_p = mgh$

Work done = potential energy gained.

Example

A 5 kg mass is lifted through a height of 2 m. $E_p = mgh = 5 \times 9.8 \times 2 = 98\,J$

KINETIC ENERGY

An **unbalanced force** applied over a **distance** causes **acceleration** and an increase in **kinetic energy**. Work is done and energy is transferred.

The gain in kinetic energy: $E_k = \frac{1}{2}mv^2$

Examples

1 An increase in speed of 0–60 mph takes four times as much energy as 0–30 mph. $E_k \propto v^2$.

2 How much more energy is required to go to 90 mph compared with going to 30 mph?
9 times!

3 How much energy is required for a 500 kg car to go to $20\,ms^{-1}$ from rest?
$E_k = \frac{1}{2}mv^2 = \frac{1}{2}500 \times 20^2 = 100\,000\,J$

CONSERVATION OF ENERGY

Energy is never created or destroyed, just changed from one form into another.

1   When a car accelerates from rest to a high speed, **work** changes to **kinetic** energy.

traffic lights

$Fd = \frac{1}{2}mv^2$ assuming no work is done against friction.

2   In real life, **friction** is **transferring energy to the surroundings**. Some work is done against friction, and the rest becomes kinetic energy.

$Fd = \frac{1}{2}mv^2 + F_{fr}\,d$

3   A ball is dropped through a height of 2 m. **Potential** energy is converted to **kinetic** energy.

$mgh = \frac{1}{2}mv^2$ assuming no air resistance.

4   A skateboarder skates from one high hill to a lower one, as shown.

> **DON'T FORGET**
>
> In E_p and E_k equations, it is only the gain or loss in energy that is calculated.

The loss in **potential** energy will be converted into **work done** against the frictional forces, and the remainder will be transferred to **kinetic** energy.

$E_p = E_W + E_k$

$mg(h_2 - h_1) = F_R\,d + \frac{1}{2}mv^2$

Note: If the skateboarder just reaches the top of hill 2, then E_k is zero.

LET'S THINK ABOUT THIS

1   When energy is transferred to the **surroundings**, it is usually in the form of **heat**. Energy is said to be degraded, as it is hard to recover this energy.

2   The force of friction is the **cause** of energy lost to the surroundings.

3   Energy conservation is a possible **alternative** to using equations of motion in some problems.

MOMENTUM AND ENERGY

MOMENTUM

Momentum measures the motion of a body. **Momentum** is the product of **mass** and **velocity**.

Momentum $p = m\,v$ Momentum has the units **kg m s^{-1}**

Momentum is a **vector** quantity. To the right, take as +ve momentum; to the left as –ve.

Conservation of momentum

The **law of conservation of linear momentum** can be applied to the **interaction** of **two objects** moving in **one dimension**, in the **absence** of net external **forces**.

Momentum is conserved. **Total momentum before = total momentum after**.

EXPLOSIONS

Consider an explosion where an object breaks into two pieces.

before   after

Total momentum before = total momentum after $mu = m_1v_1 + m_2v_2$

Example

$$mu = m_1v_1 + m_2v_2$$
$$(3 \times 0) = (2 \times -50) + (1 \times v_2)$$
$$0 = -100 + v_2$$
$$v_2 = \mathbf{100\,ms^{-1}}\ \textbf{to the right}\ (\text{since } v_2 \text{ is positive})$$

3 kg, rest   2 kg, 50 ms^{-1} 1 kg, v = ?

Energy has to be **put in** to cause the explosion. Chemical or potential energy changes to **kinetic**.

Kinetic energy before: $E_{k\ before} = \frac{1}{2}\,mu^2$ = 0 J

Kinetic energy after: $E_{k\ after} = \frac{1}{2}\,m_1v_1^2 + \frac{1}{2}\,m_2v_2^2$ $= \frac{1}{2} \times 2 \times 50^2 + \frac{1}{2} \times 1 \times 100^2 = 7500\,J$

The **kinetic energy** after is **greater** than before. Difference is the input. **E$_k$ is not conserved**.

INELASTIC COLLISIONS

Similarly to the opposite of an explosion, objects make contact and may join.

In an **inelastic** collision, **momentum is conserved** but **kinetic energy is not**.

Momentum: Total momentum before = total momentum after

$$m_1u_1 + m_2u_2 = m_1v_1 + m_2v_2$$

Energy: Energy is **given out**

Kinetic energy $\Rightarrow$ Heat + Sound

Kinetic energy before: $E_{k\ before} = \frac{1}{2}\,m_1u_1^2 + \frac{1}{2}\,m_2u_2^2$

Kinetic energy after: $E_{k\ after} = \frac{1}{2}\,m_1v_1^2 + \frac{1}{2}\,m_2v_2^2$

The **kinetic energy** after is **less than** before. **Kinetic energy is not conserved**.

ELASTIC COLLISIONS

An **elastic collision** is one where both **momentum** and **kinetic energy** are **conserved**.

Elastic collisions take place between magnets and charged particles.

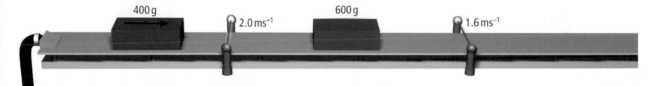

Momentum: Total momentum before = total momentum after

$$m_1u_1 + m_2u_2 = m_1v_1 + m_2v_2$$

Energy: Energy is **conserved**

Kinetic energy before: $E_{k\,before} = \frac{1}{2}\,m_1u_1^2 + \frac{1}{2}\,m_2u_2^2$

Kinetic energy after: $E_{k\,after} = \frac{1}{2}\,m_1\,v_1^2 + \frac{1}{2}\,m_2v_2^2$

If the total kinetic energies **before** and **after** are the same, then we have an **elastic collision**.

$$E_{k\,before} = E_{k\,after} \Rightarrow \frac{1}{2}\,m_1u_1^2 + \frac{1}{2}\,m_2u_2^2 = \frac{1}{2}\,m_1\,v_1^2 + \frac{1}{2}\,m_2v_2^2$$

DON'T FORGET

We can use **either** the equation for **momentum**, or for **energy**. They are **not** the same.

Example

A linear air-track vehicle, mass = 400 g and moving through a light gate at 2.0 ms⁻¹, collides with a stationary vehicle, mass = 600 g, which a second light gate records as travelling off at 1.6 ms⁻¹.

400 g   2.0 ms⁻¹   600 g   1.6 ms⁻¹

Both vehicles were fitted with magnets, and they did not appear to touch during the collision.

- Calculate the velocity of the first vehicle through one of the light gates after the collision.

- Show what type of collision this was.

Total momentum before = total momentum after

$$m_1u_1 + m_2u_2 = m_1v_1 + m_2v_2$$

$(0.4 \times 2.0) + (0.6 \times 0) = (0.4 \times v_1) + (0.6 \times 1.6)$
$v_1 = -0.4\,\text{ms}^{-1}$.
Velocity of vehicle 1 after is **0.4 ms⁻¹** to the **left**.

Kinetic energy before:

$$E_{k\,before} = \frac{1}{2}\,m_1u_1^2 + \frac{1}{2}\,m_2u_2^2 = \frac{1}{2} \times 0.4 \times 2.0^2 + 0 = 0.8\,\text{J}$$

Kinetic energy after:

$$E_{k\,after} = \frac{1}{2}\,m_1\,v_1^2 + \frac{1}{2}\,m_2v_2^2 = \frac{1}{2} \times 0.4 \times 0.4^2 + \frac{1}{2} \times 0.6 \times 1.6^2 = 0.8\,\text{J}$$

No change in kinetic energy $\Rightarrow$ **elastic collision**.

 Try http://www.fearofphysics.com/Collide/collide.html for a look at the effects of different collisions.

LET'S THINK ABOUT THIS

Collisions and explosions summary:

Momentum is conserved.
Energy is only conserved in a elastic collision; it is given out from inelastic collisions and taken in during an explosion.
Momentum is a vector.
Total energy is always conserved.

N3 AND IMPULSE

F_1 F_2

NEWTON'S THIRD LAW

Applying the **law of conservation of momentum**

To an explosion from rest
total momentum before = total momentum after
$0 = m_1v_1 + m_2v_2$
$m_1v_1 = -m_2v_2$
$\dfrac{m_1v_1}{t} = -\dfrac{m_2v_2}{t}$
$m_1a_1 = -m_2a_2$
$F_1 = -F_2$

see a) below

same contact time

Newton's Third Law;
see b) below

For a collision
total momentum before = total momentum after
$m_1u_1 + m_2u_2 = m_1v_1 + m_2v_2$
$m_1(v_1 - u_1) = -m_2(v_2 - u_2)$
$m_1\dfrac{(v_1 - u_1)}{t} = -m_2\dfrac{(v_2 - u_2)}{t}$
$m_1a_1 = -m_2a_2$
$F_1 = -F_2$

The **law of conservation of momentum** shows us that

a) the **changes in momentum** of each object are **equal in size** and **opposite in direction**.
b) the **forces** acting on each object are **equal in size** and **opposite in direction**.

IMPULSE

An object is accelerated by a force **F** for a time **t**.

$$F = ma = \frac{m(v-u)}{t} = \frac{mv - mu}{t}$$ **Unbalanced force = rate of change of momentum.**

This is how Newton first described his Second Law of motion!

Force gives the rate of change, but not the change (of momentum). Rearrange the equation further:

$$F\,t = mv - mu$$ Impulse is measured in N s or kg m s⁻¹.

- The **product F t** is called the **impulse** and is the *cause* of the change in motion.

- The **change in momentum mv – mu** is the *effect* of the impulse.

To understand the concept of impulse, we need to realise that we need to know how long a force acts for, as well as the size of the force. A force applied for 5 s causes five times more change in momentum than the same force applied for one second.

A change in momentum depends on:

1   the size of **force**, and
2   the **time** the force acts.

Example

A force of 100 N is applied to a small 150 g ball for 0.020 s. Find the final velocity.

$$F\,t = mv - mu$$

$$100 \times 0.020 = 0.150 \times (v - 0) \quad \text{velocity} = 13.3\,\text{ms}^{-1} \text{ in the direction of force.}$$

IMPULSE FROM A GRAPH

A **small force applied for a long time** causes the **same change in momentum** as a **large force applied for a short time**. This effect is used for crumple zones in cars.

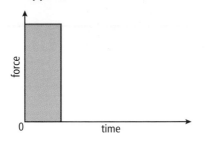

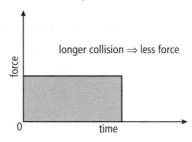

longer collision ⇒ less force

Impulse = area under a F / t graph

In real situations, the force is not usually constant. We must either

- calculate the **average force × time**, $\bar{F}t$, or

- calculate the **area** under a **F/t graph**.

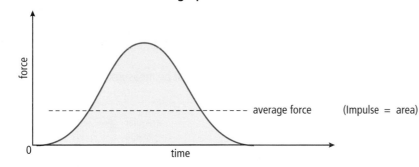

average force   (Impulse = area)

Change of momentum

Consider a ball deforming as it makes contact with a wall.

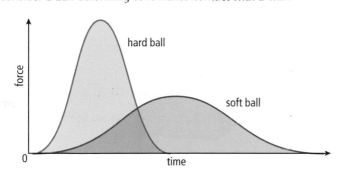
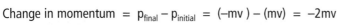

hard ball

soft ball

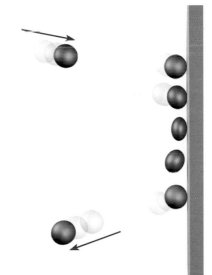

Change in momentum $= p_{final} - p_{initial} = (-mv) - (mv) = -2mv$

i.e. a change of 2mv towards the left.

LET'S THINK ABOUT THIS

What can the **law of conservation of momentum** show?
see a), b) on p. 28.

impulse = force × time

impulse = change in momentum

impulse = area under a F/t graph

DENSITY

DENSITY

The **density** of a substance is its **mass per unit volume**.

$$density = \frac{mass}{volume} \qquad \rho = \frac{m}{V} \qquad \text{Units: kg m}^{-3}$$

If $g\,cm^{-3}$ are given, this should be changed to $kg\,m^{-3}$.

Example

A laboratory has a volume of $200\,m^3$. If the density of air is $1.29\,kg\,m^{-3}$, what is the mass of the air in the laboratory?

$$\rho = \frac{m}{V} \qquad 1.29 = \frac{m}{200} \qquad m = 258\,kg$$

MEASURING DENSITY

To **calculate density** using $\rho = \frac{m}{V}$, mass and volume must be measured.

Solids

Mass is measured from a **balance**. **Volume** can be measured from **dimensions** of a regular shape or the **displacement** of water.

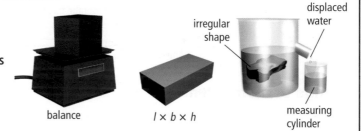

balance

$l \times b \times h$

irregular shape

displaced water

measuring cylinder

Liquids

Mass is measured from a **balance**.

Volume can be measured using a **measuring cylinder**.

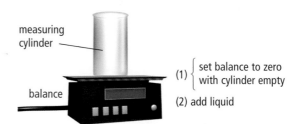

measuring cylinder

balance

(1) { set balance to zero with cylinder empty

(2) add liquid

Gases

1 Mass with gas plus flask – mass of flask = **mass** of gas.

2 The gas is then evacuated.

3 Water enters the flask to replace the gas that has been evacuated. This **volume** can be found using a measuring cylinder.

4 **Calculate** using: $\rho = \frac{m}{V}$.

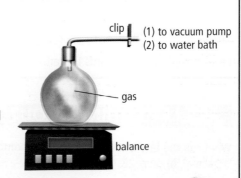

clip

(1) to vacuum pump
(2) to water bath

gas

balance

Example

Calculate the density of a 2 kg block of carbon measuring 5 cm by 10 cm by 20 cm.

$$V = 0.05 \times 0.1 \times 0.2 = 0.001\,m^3 \qquad \rho = \frac{m}{V} = \frac{2}{0.001} = 2000\,kg\,m^{-3}$$

THE STATES OF MATTER

Density

Solid	Density kg m⁻³
Copper	8960
Ice	920
Aluminium	2700
Iron	7860
Perspex	1190

Liquid	Density kg m⁻³
Sea water	1020
Water	1000
Vinegar	1050
Olive oil	915
Ethanol	791

Gas	Density kg m⁻³
Air	1.29
Steam	0.9
Oxygen	1.43
Nitrogen	1.25
Hydrogen	0.09

Solids and **liquids** have **similar** densities.

The densities of **gases** are approximately **1000 times smaller** than solids and liquids.

Density ratio	
Solids/liquids	Gases
1	10^{-3}

DON'T FORGET

You need to be able to use the **spacing** to explain the relative densities.

Spacing

Gas molecules are approximately **10 × further apart** than solid or liquid molecules.

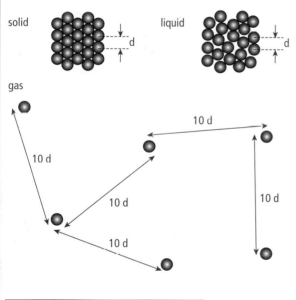

Spacing ratio	
Solids/liquids	Gases
1	10

Volumes

Gas molecules occupy approximately **1000 × more volume** (10^3) than solid or liquid molecules.

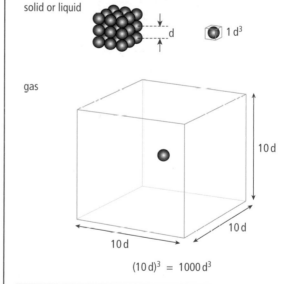

$(10\,d)^3 = 1000\,d^3$

Volume ratio	
Solids/liquids	Gases
1	10^3

LET'S THINK ABOUT THIS

1 How do you put more into a beaker once it is 'full'? If it is a gas, you can compress it by fitting a piston to increase the density!

2 Many people think that liquid molecules are much further apart than those in solids. Not true – when wax melts, it occupies a similar volume. Energy has been put in to break bonds, but it is not until the gas state that the molecules occupy much larger volumes.

PRESSURE

PRESSURE

Pressure is the **force per unit area**, when the force acts **normal** to the surface.

$$\text{pressure} = \frac{\text{force}}{\text{area}} \qquad P = \frac{F}{A}$$

Units: $N\,m^{-2}$ or pascal, Pa

1 Pascal = 1 Newton per square metre. $1\,Pa = 1\,N\,m^{-2}$

Example

A person whose mass is 60 kg stands on one foot which has an area of 200 cm². Calculate the pressure exerted on the floor.

$$200\,cm^2 = 2 \times 10^{-2}\,m^2. \qquad P = \frac{F}{A} = \frac{60 \times 9.8}{2 \times 10^{-2}} = 2.94 \times 10^4\,Pa \quad (\text{approx. } 3\,N\,cm^{-2})$$

Note: On an inclined plane, a mass exerts a force normal to the surface, equal to a component of its weight, W cos θ.

$$P = \frac{W\cos\theta}{A}$$

PRESSURE IN FLUIDS

The **pressure** at a point in a fluid (liquid or gas) at rest depends on:

* the **density**

* the **gravitational field strength** and

* the **depth**

Pressure increases with each of these quantities.

The **pressure** at a point in a liquid at rest, at depth h, is given by $P = \rho g h$

(Note that this equation does not hold for a gas, where the density increases with depth.

In a liquid, we can assume the density to be constant, as liquids are hard to compress.)

Pressure exerts a force **equally** in all directions.

Pressure is a **scalar** quantity.

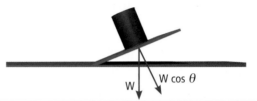

Examples

1 Calculate the pressure due to the water at a depth of 30 m in the sea. ($\rho_{\text{sea water}} = 1020\,kg\,m^{-3}$)

   $P = \rho g h = 1020 \times 9.8 \times 30 = 299\,880\,Pa.$

2 In this manometer, the pressure of the gas supply is greater than atmospheric pressure.

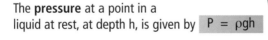

$$\begin{aligned}
P_{gas} &= P_{atm} + \rho g h \\
&= (1.10 \times 10^5) + (1000 \times 9.8 \times 0.10) \\
&= (1.10 \times 10^5) + (9.8 \times 10^2) \\
&= 110\,980\,Pa \\
&= 1.11 \times 10^5\,Pa.
\end{aligned}$$

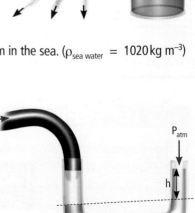

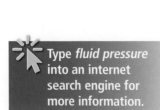
Type *fluid pressure* into an internet search engine for more information.

BUOYANCY FORCE

When an object is submerged in a fluid (liquid or gas), it **appears** to lose weight.

The weight stays the same, but the object experiences a **buoyancy force** or **upthrust**.

The buoyancy force (upthrust) can be explained in terms of the pressure difference between the top and bottom of an object.

The **pressure** at the bottom is greater than the pressure at the top.

$$\rho gh_{bottom} > \rho gh_{top}$$

The top area equals the bottom area.

$$F = PA \qquad P_{bottom} A > P_{top} A$$

The **force** at the bottom upwards is greater than the force at the top downwards.

The **difference** in these forces is the **upthrust** or **buoyancy force**.

Once we have calculated upthrust, we can compare it with the weight:

If **weight > upthrust**, then the object will sink.

If **weight < upthrust**, then the object will rise.

If **weight = upthrust**, then the object will float.

upthrust

W = mg

Look up *buoyancy* at http://www.exploratorium.edu/ for examples.

DON'T FORGET

Upthrust stays the **same** even at different depths, as the **difference** in depth is not changed.
A **submarine** will take in or release water to **change its weight** to sink or rise.

LET'S THINK ABOUT THIS

1 Increased loads require increased depths to be supported.

$\Rightarrow$ Pressure increases with depth.

different loads

2 An increased depth is required to support the same load in a less dense solution.

$\Rightarrow$ Pressure increases with density.

different densities

3 To raise a **sunken ship**, its **weight needs to become less** than the upthrust.

- Fill the ship with polystyrene or air to displace water.
- The density of polystyrene or air is less than the water.
- The weight of the ship has been reduced.
- Weight < upthrust.
- The ship rises.

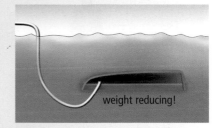

weight reducing!

4 The apparent loss in weight = upthrust.

5 The apparent loss in weight = weight of water displaced. This is the Archimedes Principle. Eureka!

displaced water

GAS LAWS 1 – PRESSURE LAW AND CHARLES' LAW

KINETIC MODEL OF A GAS

Gases are made up of **tiny particles**, freely moving in **random directions** at **high speeds**.

Gas particles make **elastic collisions** with each other and the walls of their container. Gas particles are moving with an **average kinetic energy** that is proportional to the temperature of the gas.

The **collisions** between gas particles create tiny **forces** on the walls of the container, and this creates pressure.

The **pressure** of a gas can be measured with a **Bourdon gauge**.

Atmospheric pressure $= 1.01 \times 10^5$ Pa.

The **gas laws** give the relationship between the **pressure**, **temperature** and **volume** of a **fixed mass** of gas. **Kinetic theory** then explains the relationship in terms of **movement** of the **particles**.

DON'T FORGET

You need to explain how the kinetic model accounts for the pressure of a gas.

PRESSURE AND TEMPERATURE

Consider the relationship between **pressure** and **temperature** of a **fixed mass** and **fixed volume** of gas. The water is heated slowly, so the gas is the same temperature as the water.

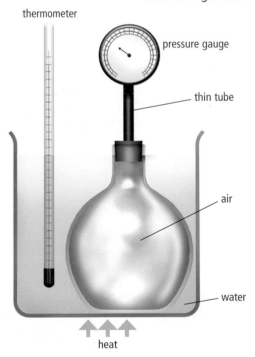

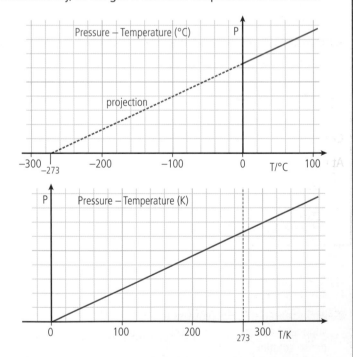

Zero pressure appears at −273°C.

Zero pressure indicates the **true zero** of temperature, called **Absolute Zero**.

Absolute zero is 0 K, and the **Kelvin scale** uses the same size of divisions as the Celsius scale.

On the Kelvin scale, **pressure** is **directly proportional** to **temperature**: P ∝ T

DON'T FORGET

You must change degrees Celsius to kelvin.

$$\frac{P}{T} = k \qquad \frac{P_1}{T_1} = \frac{P_2}{T_2}$$ This is the **pressure law**.

VOLUME AND TEMPERATURE

Consider the relationship between **volume** and **temperature** of a **fixed mass** of gas at a **constant pressure**.

The water is heated slowly, so the gas is the same temperature as the water.

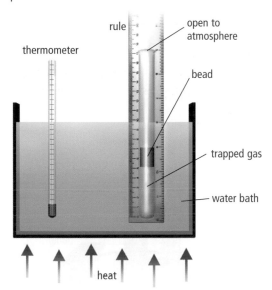

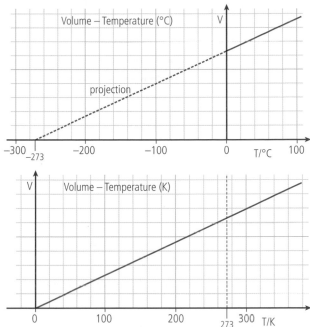

The bead moves so that the pressure of trapped gas stays in equilibrium with atmospheric pressure.

Zero volume appears at −273°C.

Zero volume indicates the **true zero** of temperature, called **Absolute Zero**.

Absolute zero is 0 K, and the **Kelvin scale** uses the same size of divisions as the Celsius scale.

On the Kelvin scale, **volume** is **directly proportional** to **temperature**: $V \propto T$

$$\frac{V}{T} = k \qquad \frac{V_1}{T_1} = \frac{V_2}{T_2}$$ This is **Charles' Law**.

> **DON'T FORGET**
> You must change degrees Celsius to kelvin.

Example

At what temperature will a litre of trapped gas have its volume doubled if it was at −20°C?

−20 °C = 253 K.

$$\frac{V_1}{T_1} = \frac{V_2}{T_2} \Rightarrow \frac{1}{253} = \frac{2}{T_2} \Rightarrow T_2 = 506\,\text{K. New temperature} = 233°\text{C}.$$

TEMPERATURE SCALES

degree Celsius to kelvin: add 273

Only the **kelvin scale** starts from an **absolute zero** of temperature.

kelvin to degree Celsius: subtract 273

> **DON'T FORGET**
> A temperature change of 100°C = a temperature change of 100 K. The **divisions** are the **same size**.

LET'S THINK ABOUT THIS

1 The only true temperature scale is the Kelvin scale. There should be no such thing as negative temperatures, but the Celsius scale has a more useful range for everyday purposes.

2 What is twice the temperature of 20°C? $2 \times 293 = 586\,\text{K} = 313°\text{C}$

GAS LAWS 2 – BOYLE'S LAW AND PARTICLE THEORY

PRESSURE AND VOLUME

Consider the relationship between **pressure** and **volume** of a **fixed mass** of gas at a **constant temperature**.

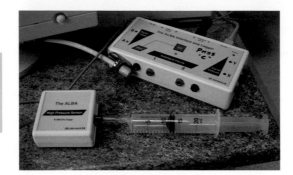

As the pressure increases, the volume decreases.

The **pressure** is **inversely proportional** to the **volume**.

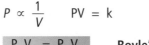

$$P \propto \frac{1}{V} \qquad PV = k$$

$$P_1 V_1 = P_2 V_2 \qquad \textbf{Boyle's Law}$$

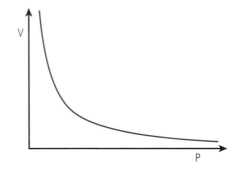

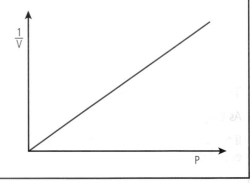

THE GENERAL GAS EQUATION

The three gas laws can be combined into one general gas equation. P, V, and T can all change.

1 $\dfrac{P_1}{T_1} = \dfrac{P_2}{T_2}$

2 $\dfrac{V_1}{T_1} = \dfrac{V_2}{T_2}$ } Remember temperatures in Kelvin!

3 $P_1 V_1 = P_2 V_2$

The **general gas equation**: $\dfrac{P_1 V_1}{T_1} = \dfrac{P_2 V_2}{T_2}$

The general equation can be used in all gas calculations.

A constant quantity will simply cancel, e.g. $T_1 = T_2 \Rightarrow$ equation 3.

GAS LAWS AND THE KINETIC MODEL

Pressure, **temperature** and **volume** can be explained in terms of the **movement** of the **particles** of the gas. The particles move in random directions.

Pressure: The **pressure** exerted by a gas is caused by the **moving particles** colliding with the **walls** of the container. Each **collision** causes the wall to receive a **tiny force**. The pressure depends on the **number of collisions** and the **average force per collision** exerted on the **area** of the walls.

Temperature: **Heat** is **energy**, which, when added to a gas, increases both the **velocity** and the **kinetic energy** of the particles. (**Temperature** is proportional to the **average kinetic energy** of the **particles** but not the velocity.)

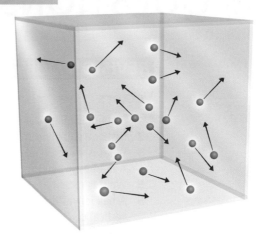

Volume: The **volume** of a gas is the **amount of space** the **particles** have to move around in. If the walls of the container are brought closer, the particles take **less time** between **collisions** with the walls.

The pressure–temperature law (the pressure law)

If the **temperature increases**, the **particles** have **more kinetic energy** and the **velocity increases**. The **particles** hit the container walls **more often** and with **greater force**.

As the pressure **P** increases: $P = \dfrac{F}{A}$ **A** remains constant so **F** increases.

The volume–temperature law (Charles' Law)

If the **temperature increases**, the **particles** have **more kinetic energy** and the **velocity increases**. The **particles** hit the container walls with **more force** but **less often** as the volume increases to keep the pressure constant. If the volume increases, the surface area also has to increase.

The pressure **P** is constant: $P = \dfrac{F}{A}$ **F** increases and **A** increases.

The pressure–volume law (Boyle's Law)

As the temperature is constant, the kinetic energy and the velocity are constant.

If the **volume** is **increased**, the **particles** hit the container walls **less often**, therefore they exert **less average force** on the **increased area** of the container walls.

The pressure **P** decreases: $P = \dfrac{F}{A}$ **F** decreases and **A** increases.

DON'T FORGET

The kinetic model uses the **motion** of the **particles** to explain the gas laws.

LET'S THINK ABOUT THIS

The general equation is the most useful of the gas equations, as it gives you **four in one**: the general equation, the pressure law, Charles' Law and Boyle's Law.

KEY QUESTIONS

REVISION QUESTIONS

Vectors

1 Distinguish between distance and displacement.
2 Distinguish between speed and velocity.
3 What is a vector quantity?
4 A man walks 300 m east then 200 m south. What is the man's displacement?
5 What is meant by the resultant of a number of forces?
6 A ball leaves the ground with a velocity of 40 ms⁻¹ at 50° from the horizontal. What is the vertical component of this velocity?

Equations of motion

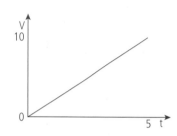

7 What is acceleration?
8 Describe the principles of a method for measuring acceleration.
9 From this velocity-time graph, draw the corresponding acceleration-time graph.
10 Label both graphs to describe the motion appropriately.
11 Derive the equations of motion from basic definitions.
12 What is the final speed of a car which accelerates from rest at 5 ms⁻² for a distance of 20 m?

Newton, energy and power

13 Define the newton.
14 A toy rocket of mass 2 kg takes off with a thrust of 300 N. Calculate the initial acceleration.
15 A skier of mass 80 kg descends 100 m down a slope. If her speed at the bottom is 30 ms⁻¹, what was the average frictional force?

Momentum and impulse

16 What is momentum?
17 How is the law of conservation of linear momentum applied?
18 What is an elastic collision?
19 What is an inelastic collision?
20 A 6 kg vehicle moving at 2 ms⁻¹ collides elastically with a stationary 3 kg vehicle. What is the new velocity of the stationary vehicle if the first vehicle is slowed to 1 ms⁻¹?
21 A gun of mass 100 kg fires a bullet of mass 1 kg with a velocity of 200 ms⁻¹. Calculate the gun's recoil velocity.
22 What two items can the law of conservation of momentum show?
23 Write a relationship for impulse based on its cause.
24 Write a relationship for impulse based on its effect.
25 Write two units for impulse.
26 A 800 kg car crashes into a wall at 20 ms⁻¹. If the contact time was 400 ms calculate the change in momentum of the car and the average force exerted.

Density and pressure

27 Define density.
28 If a cube of length 3 cm has a mass of 45 g. What is its density in SI units?
29 Describe the principles of a method of measuring the density of air.
30 State and explain the relative magnitudes of the densities of solids, liquids and gases.
31 Define pressure.
32 What is the equivalent of one newton per square metre?
33 A block of mass 36 kg has dimensions 2 m by 1 m by 50 cm. What is the least pressure it can exert on a flat horizontal surface?
34 What does the pressure at a point in a fluid at rest depend on?
35 What is the pressure due to a depth of water of 2 m?
36 Explain buoyancy force or upthrust.

contd

REVISION QUESTIONS contd

Gas laws

37 Describe how the kinetic model accounts for the pressure of a gas.

38 Describe the relationship between pressure and volume of a gas.

39 Describe the relationship between pressure and temperature of a gas.

40 Describe the relationship between volume and temperature of a gas.

41 How do you convert °C to K?

42 What is 400 K in °C?

43 A gas bubble has a volume of 10 cm^3 at room pressure and normal temperature (1 × 10^5 Pa, 20°C). What is the new temperature if the pressure increases by half again and the volume is now 20 cm^3.

44 What is meant by the absolute zero of temperature?

45 Explain each of the three gas laws qualitatively in terms of a kinetic model.

ANSWERS

1 Displacement is the distance travelled in a stated direction from the starting point.

2 Speed is the distance travelled in unit time. Velocity is the displacement per unit time.

3 A vector quantity is defined by both its magnitude and its direction.

4 361 m @ 34° S of E.

5 The single force, which has the same effect as the sum of the individual forces.

6 31 ms^{-1}. **7** Acceleration is the rate of change of velocity. **8** See p 14.

9 As graph.

10 Constantly increasing velocity, constant acceleration.

11 See p 16. **12** 14 ms^{-1}.

13 The unit of force is the newton. 1 N is defined as the resultant force, which causes a mass of 1 kg to accelerate at 1 ms^{-2}.

14 140.2 ms^{-2} ⇒ 140 ms^{-2} (sig figs). **15** 424 N.

16 Momentum is the product of mass and velocity.

17 The law of conservation of linear momentum can be applied to the interaction of two objects moving in one direction, in the absence of net external forces.

18 An elastic collision is one where both momentum and kinetic energy are conserved.

19 In an inelastic collision momentum is conserved but kinetic energy is not.

20 2 ms^{-1} in original direction.

21 2 ms^{-1} in opposite direction to bullet.

22 The law of conservation of momentum applied to two objects moving in one direction shows us that: **(a)** The changes in momentum of each object are equal in size and opposite in direction. **(b)** The forces acting on each object are equal in size and opposite in direction.

23 Impulse = force × time. **24** Impulse = change of momentum.

25 N s and kg m s^{-2} **26** 16000 kg m s^{-2} and 40 000 N

27 Density is mass per unit volume. **28** 1667 kg m^{-3}. **29** See p 30. **30** See p 31.

31 Pressure is the force per unit area, when the force acts normal to the surface.

32 One pascal. **33** 176 Pa.

34 The pressure at a point in a fluid (liquid or gas) at rest depends on the density, the gravitational field strength and, the depth.

35 19600 Pa. **36** See p 33. **37** See p 34.

38 The pressure of a fixed mass of gas at constant temperature is inversely proportional to its volume.

39 The pressure of a fixed mass of gas at constant volume is directly proportional to its temperature measured in kelvins (K).

40 The volume of a fixed mass of gas at constant pressure is directly proportional to its temperature measured in kelvins (K).

41 Add 273. **42** 127 °C. **43** 606 °C.

44 Absolute zero is the true zero of temperature where volume and pressure are zero.

45 See p 37.

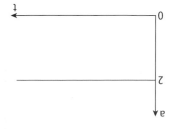

ELECTRIC FIELDS AND POTENTIAL DIFFERENCE

ELECTRIC FIELDS

There are two types of charge: **positive** and **negative**.

Like charges repel   Unlike charges attract

The region round an electric charge is an **electric field**.

In an **electric field**, an **electric charge** experiences a **force**.

The **electric field strength** is a measure of the **force** on a **unit charge**.

It sometimes helps to compare **electric fields** with **gravitational fields**:

In a **gravitational field**, a **mass** experiences a **force**.

The **gravitational field strength** is a measure of the **force** on a **unit mass**.

Gravitational field lines show the **downward direction** that **mass** will move in the field due to the force on it.

Electric field lines are designed to show the direction that a **free positive charge** will move in an **electric field**. The **direction of a field** round a charge depends on whether its sign is **positive** or **negative**.

The **electric field** causes the **free charges** in it to **move**.

Radial fields

Note that the field strength decreases with distance.

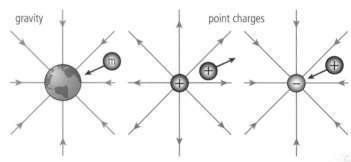

Uniform fields

Note that the field strength is constant.

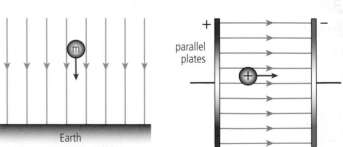

Combining fields

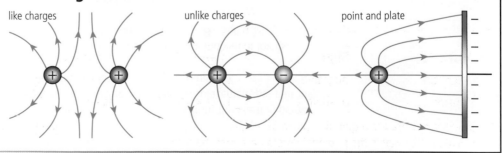

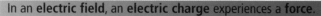

WORK DONE

When **work is done** to push a charge against a field, the charge has a gain in potential energy.

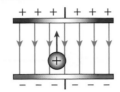

When the field **does work** on the charge, the charge accelerates and gains kinetic energy.

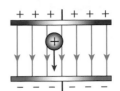

Work (W) is done when a **charge (Q)** is **moved** in an **electric field**.

The amount of work done depends on:
- the size of the charge being moved, **Q**
- the size of the voltage **V** creating the electric field.

$$W = QV \quad \text{or} \quad E_W = QV$$

A charge accelerated parallel to a field will gain **kinetic energy** from the **work done** by the field.

$$QV = \tfrac{1}{2}mv^2$$

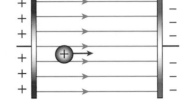

Example

An electron is accelerated by a voltage of 2000 V. What is its final speed?

$$QV = \tfrac{1}{2}mv^2 \quad v = \sqrt{\frac{2QV}{m}} \quad v = \sqrt{\frac{2 \times 1.6 \times 10^{-19} \times 2000}{9.11 \times 10^{-31}}} \quad v = 2.65 \times 10^7 \text{ m s}^{-1}$$

Potential difference

$V = \dfrac{W}{Q}$

The **potential difference (p.d.)** measured in volts (V) between two points is a measure of the **work done (W)** in moving **one coulomb of charge (Q)** between the two points.

If **1 joule** of **work** is done moving **one coulomb (1 C)** of **charge** between two points, the **potential difference** between the two points is **one volt**.

Examples

1    $3V = 3 JC^{-1}$ (3 J of work for each 1 C of charge)
2    $900V = 900 JC^{-1}$ (900 J of work for each 1 C of charge)

ELECTRIC FIELDS ACROSS A CONDUCTOR

An **electric field** applied to a **conductor** causes the free electric **charges** in it to **move**.

A 12 V battery supplies 12 J of work to each 1 C of charge.

These equations also apply to electric circuits:

 $W = QV$ $V = \dfrac{W}{Q}$

LET'S THINK ABOUT THIS

1   Useful data: check the data page in the exam!

Particle	Sign	Charge (Q)	Mass (m)
electron	negative –	1.60×10^{-19} C	9.11×10^{-31} kg
proton	positive +	1.60×10^{-19} C	1.673×10^{-27} kg

2   If 6000 J of work are required to move 3 C of charge through an electric field, what is the potential difference across the field? $V = \dfrac{W}{Q} = \dfrac{6000}{3} = 2000\,V$

ELECTRICITY AND ELECTRONICS

CIRCUIT THEORY

BASIC CIRCUIT THEORY

DON'T FORGET

Some **circuit theory** you will be familiar with, some will be new. Practise it all with more complex examples until the **problem-solving** skills are well developed.

Current

Current (I) is the **rate** of flow of **charge (Q)**.

Current is the amount of **charge** in **unit time** (1s).

$$I = \frac{Q}{t}$$ $$Q = It$$ Current is measured in **amperes (A)**.

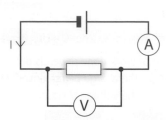

Potential difference (voltage)

$$V = \frac{W}{Q}$$ The **potential difference (p.d.)** between two points is a measure of the **work done** in moving **one coulomb of charge** between the two points, e.g. across a resistor.

Voltage is measured in volts (**V**).

$$V = \frac{E}{Q}$$ In a resistor, the work done becomes heat, and the **voltage** is also a measure of the **energy** given out per **unit charge** (1 C).

Work is done to move charges through the components of a circuit. The **work done** comes from the **energy supplied** to the charges as they pass through the **source**.

Electromotive force (e.m.f.)

The **e.m.f.** of a source is the **electrical potential energy** supplied to each **coulomb of charge** which passes through the **source**.

E.m.f. is measured in volts (J C^{-1}).

Work or energy

Work or energy is measured in **joules (J)**.

$$W = QV$$ or $$E_W = QV$$ Can combine with $Q = It$ $$E_W = ItV$$

Resistance

Resistance is measured in **ohms (Ω)**.

Ohm's Law $$R = \frac{V}{I}$$

$\frac{V}{I}$ is a **constant** for most resistors.

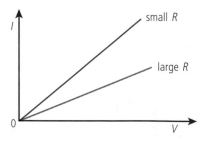

Power

Power is the **rate** of doing **work**.

Power is the amount of **work done** in **unit time** (1 s).

Power is measured in **watts (W)**. $$P = \frac{E}{t}$$

$$P = \frac{E}{t} = \frac{ItV}{t} = IV$$ $$P = IV$$

$$P = IV = I(IR) = I^2R$$ $$P = I^2R$$

$$P = IV = \left(\frac{V}{R}\right)V = \frac{V^2}{R}$$ $$P = \frac{V^2}{R}$$

TOTAL RESISTANCE

Conservation of energy

From **conservation of energy**, we know that energy is not created or destroyed. The energy supplied per coulomb to the charges as they pass through the source must **equal** the energy dissipated per coulomb by the charges in the circuit.

The **sum** of the **e.m.f.**s round a circuit is **equal** to the **sum** of the **p.d.**s round the circuit.

$$\Sigma E = \Sigma IR \qquad \text{or} \qquad \Sigma E = V_1 + V_2 + V_3$$

$E = 3\,V$

$E = 3\,V$

$V_1 = 2\,V$

$V_2 = 2\,V$

$V_2 = 2\,V$

6 V   6 V

Resistors in series

From conservation of energy:

$$E = V_1 + V_2 + V_3$$

I constant

$$IR_T = IR_1 + IR_2 + IR_3$$
$$IR_T = I(R_1 + R_2 + R_3)$$
$$R_T = R_1 + R_2 + R_3$$

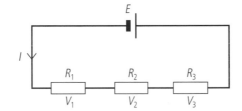

DON'T FORGET

Resistor derivations need to be learned.

Resistors in parallel

From conservation of charge:

$$I = I_1 + I_2 + I_3$$
$$\frac{V}{R_T} = \frac{V}{R_1} + \frac{V}{R_2} + \frac{V}{R_3}$$

V constant

$$V\frac{1}{R_T} = V\left(\frac{1}{R_1} + \frac{1}{R_2} + \frac{1}{R_3}\right)$$
$$\frac{1}{R_T} = \frac{1}{R_1} + \frac{1}{R_2} + \frac{1}{R_3}$$

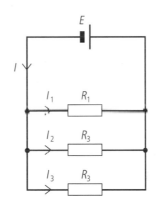

a resistor

LET'S THINK ABOUT THIS

1 Series and parallel circuits

Series:
$$E = V_1 + V_2 + V_3$$
$$I = I_1 = I_2 = I_3$$
$$R_T = R_1 + R_2 + R_3$$

Parallel:
$$E = V_1 = V_2 = V_3$$
$$I = I_1 + I_2 + I_3$$
$$\frac{1}{R_T} = \frac{1}{R_1} + \frac{1}{R_2} + \frac{1}{R_3}$$

2 Combining $4\,\Omega$, $6\,\Omega$ and $8\,\Omega$ resistors.

Resistors in **series increase** the total resistance.

$$R_T = R_1 + R_2 = R_3 = 4 + 6 + 8 = 18\Omega$$

Resistors in **parallel decrease** the total resistance.

The total resistance is always smaller than the smallest resistor.

$$\frac{1}{R_T} = \frac{1}{R_1} + \frac{1}{R_2} + \frac{1}{R_3} = \frac{1}{4} + \frac{1}{6} + \frac{1}{8} = \frac{13}{24} \Rightarrow R_T = \frac{24}{13} = 1.8\Omega$$

INTERNAL RESISTANCE

INTERNAL RESISTANCE OF A POWER SOURCE

The simple **cell** is used in many circuits. Most cells show the **e.m.f.** (the energy supplied per coulomb of charge). Yet, when we use it, the voltage available (called the **terminal potential difference** or t.p.d.) is less than the stated voltage! Why?

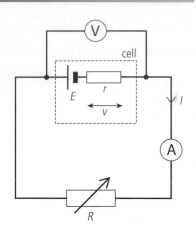

In fact, the **potential difference** across the **terminals** of a cell, a battery of cells, or any source, **decreases** as the **current drawn** from the source **increases**.

When electrons travel through a source, they have to overcome some resistance. Some of the **energy** from the e.m.f. is used up inside the cell by the **charges** crossing this **internal resistance**.

An **electrical source** is **equivalent** to a source of **e.m.f.** with a resistor in series, the **internal resistance (r)**.

cell has internal resistance

The **voltage dropped** across the **internal** resistance of a cell is often called the **lost volts (v)**.

The voltage available at the terminals is called the **terminal potential difference** or **t.p.d. (V)**.

E = e.m.f. of the source.
V = potential difference across the terminals (t.p.d.).
R = resistance of the external circuit.
I = current drawn from the source.
r = internal resistance of the source.
v = the lost volts.

Alternative equations

E	=	V	+	v	the e.m.f. is shared
e.m.f.		t.p.d.		lost volts	

E	=	IR	+	Ir	$V = IR$ and $v = Ir$

V	=	E	−	Ir	t.p.d. = e.m.f. minus the lost volts

$$I = \frac{E}{R+r} \qquad \text{current}$$

$$r = \frac{v}{I} = \frac{E-V}{I} \qquad \text{internal resistance}$$

As **circuit resistance R decreases**, the **current I increases** and the **lost volts v** across the internal resistance **increases**. Thus the **terminal potential difference V decreases**.

Example

A 9V cell has an internal resistance of 0.5 Ω. What is the t.p.d. when the current drawn is 3A?

$V = E - Ir = 9 - (3 \times 0.5) = 7.5\text{V}$.

DON'T FORGET

The internal resistance **r** of a cell is considered to be a constant.

MEASURING E.M.F. AND INTERNAL RESISTANCE

The **e.m.f.** of a source can be measured using a **voltmeter** with high resistance as negligible current is drawn.

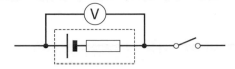

When **no current** is drawn, there are no lost volts, so the t.p.d. is equal to the **e.m.f.**

The **e.m.f.** and **internal resistance** can be found from the following circuits.

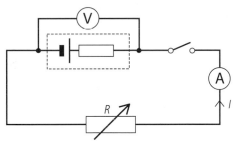

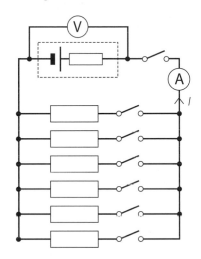

- Measure I and V for various values of load resistance.

- Plot a graph of voltage against current.

When the **load resistance is decreased** (**R** reduced or more parallel branches added), the **current will increase** and **t.p.d. will decrease**.

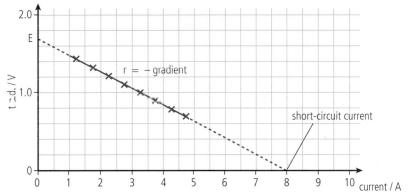

For more information search *internal resistance* at www.iop.org.

| E = the y intercept | e.m.f. is the voltage when the **current is zero**. |
| r = – the gradient | **internal resistance** from the **gradient** of the graph. |

The **intercept** on the **current axis** gives the maximum 'short-circuit current', which has been limited by the internal resistance.

DON'T FORGET

When doing this experiment, you cannot just take one pair of readings and then use an equation, due to the **uncertainties in measurement**.

LET'S THINK ABOUT THIS

1 For the mathematical:

$$E = V + Ir$$

$$\Rightarrow V = -rI + E$$

$$y = mx + c \qquad c = E \quad and \quad m = -r$$

2 r is constant. When R changes, the ratio r:R changes so the ratio v:V also changes.

3 V = IR should only be used with the external resistance. The whole-circuit equivalent is:

$$E = I(R + r) \qquad or \qquad E = IR + Ir.$$

THE WHEATSTONE BRIDGE

THE WHEATSTONE BRIDGE

The **Wheatstone bridge circuit** is used in two ways:

1 The **balanced circuit** is used to find the value of an **unknown resistor** by comparison to standard resistors.

2 The **out-of-balance circuit** is used to detect small **changes** in a **resistance transducer** (thermistor, LDR etc.).

The Wheatstone bridge is made of two potential dividers in parallel.

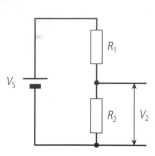

Potential divider circuit

Potential divider equations: $V_2 = \dfrac{R_2}{R_1 + R_2} V_s$ or $\dfrac{V_1}{V_2} = \dfrac{R_1}{R_2}$

In the Wheatstone bridge, a sensitive voltmeter bridges the outputs of the potential dividers.

BALANCED CIRCUIT

When the **Wheatstone bridge** is **balanced**, the **potential difference** across the voltmeter is **zero**. The **potential** at each side of the voltmeter is the **same**. **Balance** occurs when the **ratio** of the **resistors** in each **potential divider** is the **same**.

$V = 0$ volts.

$\dfrac{R_1}{R_2} = \dfrac{R_3}{R_4}$

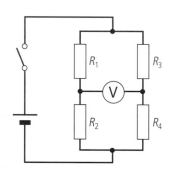

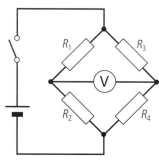

Modern layout　　　　　　　*Traditional layout*

DON'T FORGET

The balance condition does **not** depend on the supply voltage.

R_1 and R_2 are normally **fixed-value** high-quality resistors.

R_3 is often replaced with a **decade resistance box** allowing a choice of resistance.

R_4 is then an **unknown resistor** whose value is to be found.

Example

In a Wheatstone bridge circuit, $R_1 = 100\ \Omega$, $R_2 = 1000\ \Omega$. The voltmeter reads 0 V when the decade resistance box is set at 470 Ω. What is the value of the fourth resistor?

$\dfrac{R_1}{R_2} = \dfrac{R_3}{R_4}$　　　$\dfrac{100}{1000} = \dfrac{470}{R_4}$　　　$R_4 = \dfrac{470 \times 1000}{100} = 4700\ \Omega$

Note: An advantage of this circuit is that it does not depend on obtaining accurate readings from ammeters or voltmeters.

Sometimes the voltmeter has a resistor in series with it to protect it from possible large currents in the early adjustments. This is then bypassed to obtain a more sensitive voltmeter.

OUT-OF-BALANCE CIRCUIT

A **Wheatstone bridge** can be used to measure **small changes** in **resistance**.

1   The bridge is **initially balanced**. Alter variable resistor R_v until $V = 0V$.

2   The value of **one** of the resistors is **changed** by a small amount.

3   A **voltage** is noted on the **voltmeter**.

4   **Increase** the **change** in **resistance**.

5   A **larger voltage** appears on the **voltmeter**.

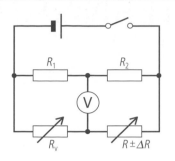

If the value of one resistor is changed by a small amount in an **initially balanced** Wheatstone bridge, the **out-of-balance p.d.** is **directly proportional** to the **change in resistance**.

$$V \propto \Delta R$$

By replacing the $R \pm \Delta R$ resistor block with a resistance transducer, many **measurement instruments** can be made.

A resistance transducer's **resistance** will **change** with a physical property like **temperature**, **light** or **force**.

Applications

Application	Resistance transducer	Sensor images
Temperature monitoring	Thermistor or platinum film resistor	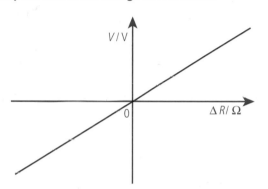 bead thermistor, rod thermistor, disc thermistor, thermistor circuit symbol; connection to leads, sheath, resistance thermometer, connection leads, insulator
Gas flow meter	Thermistor or platinum film resistor	Gas flow cools the thermometer.
Light meter	LDR Light-dependent resistor	
Strain measurement	Strain gauge	

Any **changes** in the transducer's **resistance** show up as a change in detector **voltage** as it varies **from zero**.

LET'S THINK ABOUT THIS

1   The voltmeter scale will not be left in volts. It will be **calibrated** to read **temperature**, **light**, **force** or any **physical quantity** being measured.

2   The voltmeter can be replaced with a computer interface and **voltage sensor**.

ALTERNATING CURRENT AND VOLTAGE

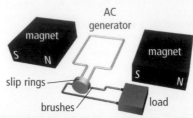

ALTERNATING CURRENT A.C.

In a power station, a generator has conductors which move up and down through a magnetic field as they rotate. **Current** and **voltage** are induced in **alternating directions**. These alternating current (a.c.) supplies are then sent to our homes using transformers.

SIGNALS ON THE OSCILLOSCOPE

The **cathode ray oscilloscope** (CRO) has two axes called **x** and **y** – just like in maths!

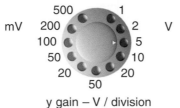

y gain – V / division

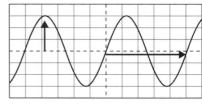

time base – s / division

Y-gain: the **y-gain** changes the scale on the y-axis.
The y-axis measures **voltage** in V cm^{-1} or V div^{-1}.

Time base: the **x-gain** changes the scale on the x-axis.
The x-axis measures **time** in s cm^{-1} or s div^{-1}.

The **voltage** alternates in a **sinusoidal** wave with a frequency **f**.
The frequency can be calculated from the **period** of time for one wave.

Example

$$f = \frac{1}{T}$$ On the screen, the period T $= 4 \times 10 = 40\,ms = 0.040\,s$

$$f = \frac{1}{T} = \frac{1}{0.040} = 25\,Hz$$ Peak voltage $= 3 \times 5 = 15V$

THE MAINS SUPPLY

The mains electricity supply has a **frequency f** $= $ **50 Hz**.

The **period** of the mains supply: $T = \frac{1}{f}$

$$T = \frac{1}{f} = \frac{1}{50} = 0.02\,s$$ **T $= $ 0.02 s**

The effective **voltage** $= $ **230V**

The peak voltage $= $ 325V approx.

PEAK AND R.M.S. VALUES

The **peak voltage** (from zero or rest to the peak) is the **maximum** reached by a supply at only two points in each cycle. The peak voltage can be found using an oscilloscope.

The **quoted voltage** is also called the **effective** or **r.m.s.** voltage.

The **quoted voltage** is the value of an a.c. voltage, which will provide the **same amount** of **energy** as a d.c. voltage.

The relationship between **peak and r.m.s. voltage** is:

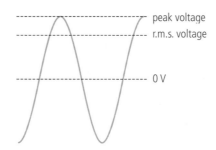

peak voltage
r.m.s. voltage

0 V

$$V_{rms} = \frac{V_{peak}}{\sqrt{2}} \quad \text{or} \quad V_{peak} = \sqrt{2}\, V_{rms}$$

Current alternates in a sinusoidal wave like voltage.

Current has a **peak** and an **effective** or **r.m.s.** value.

The relationship between peak and r.m.s. current is:

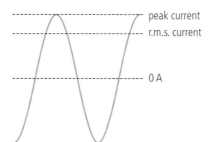

peak current
r.m.s. current

0 A

$$I_{rms} = \frac{I_{peak}}{\sqrt{2}} \quad \text{or} \quad I_{peak} = \sqrt{2}\, I_{rms}$$

Examples

1   The dial on an a.c. power supply shows 10 V. What is the peak voltage?

$$V_{peak} = \sqrt{2}\, V_{rms} = 10\sqrt{2} = 14.1\ V$$

2   If the peak voltage is 10 V, what is the r.m.s. voltage?

$$V_{rms} = \frac{V_{peak}}{\sqrt{2}} = \frac{10}{\sqrt{2}} = 7.1\ V$$

3   An a.c. ammeter reads 5 A. What is the peak current?

$$I_{peak} = \sqrt{2}\, I_{rms} = 5\sqrt{2} = 7.1\ A$$

Ohm's Law:
Ohm's Law applies for an a.c. waveform

$$V_{peak} = I_{peak}\, R$$
$$V_{rms} = I_{rms}\, R$$

DON'T FORGET

The **effective** value is always **less** than the **peak** value.
The value quoted for **a.c. power supplies** is the effective or r.m.s. value.
a.c. ammeters and **voltmeters** are calibrated to give the **effective** or **r.m.s.** value.

FREQUENCY IN A RESISTIVE CIRCUIT

The frequency of the signal generator is varied.

The **current** through a **resistor** is **unaffected** by changes of **frequency.**

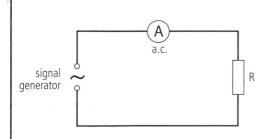

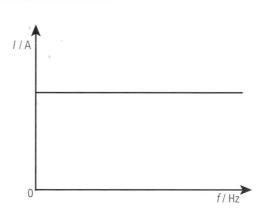

LET'S THINK ABOUT THIS

1   In practice, it is easier to measure the peak voltage by measuring the peak–peak voltage then taking half.

2   r.m.s. = root mean square. This comes from the mathematical derivation of the equations.

CAPACITANCE

STORING CHARGE

Charge can be **stored** on **parallel metal plates** by temporarily connecting them to a **direct current (d.c.)** source.

Electrons leave one plate and at the same time electrons are added to the other plate.

The **energy** to cause this **transfer of charge** from one plate to the other is the **work done** by the source.

The plates build up charge until the **potential difference** which is created across the **plates** is **equal** in size to the **potential difference** of the **source**.

The **amount of charge** that can be **stored** depends on the strength or **potential difference** of the source.

The **charge Q** on two parallel conducting plates is **directly proportional** to the **p.d. V** between the plates.

$Q \propto V$

The parallel plates **store** the **energy** supplied to them in an **electric field** between the plates.

When the source is disconnected, the **charge** and **energy** are **stored**.

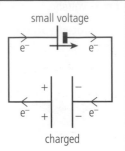

small voltage

charged

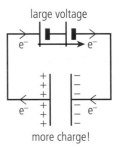

large voltage

more charge!

CAPACITANCE

A **capacitor** is a component that stores charge.

Most parallel-plate designs are rolled into a small cylinder.

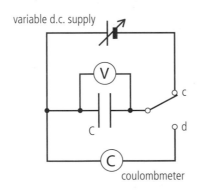
variable d.c. supply

coulombmeter

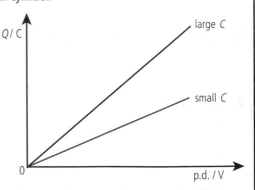

Q/C
large C
small C
0
p.d. / V

The capacitor is switched to the source to charge it. The p.d. **V** is noted. The capacitor is then switched to discharge through the coulombmeter, and the amount of charge **Q** collected is noted. The experiment is repeated over a range of supply voltages, and results are graphed.

The **charge** on a **capacitor** is **directly proportional** to the **p.d.** across the capacitor.

Capacitance is the **ratio** of **charge** to **p.d.** $C = \dfrac{Q}{V}$ $Q = CV$

Capacitance is the amount of **charge** stored **per volt**.

The unit of capacitance is the farad, and one farad is one coulomb per volt.

Most capacitors have small values:

microfarad μF 10^{-6}F   nanofarad nF 10^{-9}F   picofarad pF 10^{-12}F

DON'T FORGET

A 2V battery will store twice as much charge on a capacitor as a 1V battery.

ENERGY STORED IN A CAPACITOR

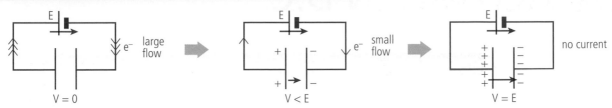

V = 0 | large flow | V < E | small flow | V = E | no current

1   Initially, when a **capacitor** is connected to a source, there is a **large flow of electrons** off one plate and onto the other.

2   As **charge builds up** on the plates (one side negative, one side positive), they create an electrostatic **force** that **opposes the flow**. The flow **rate** reduces.

3   A **potential difference V** has been created across the capacitor in **opposition** to the **source's e.m.f. E**. However, as E is larger than V, there is still a flow of electrons.

4   Finally, the **potential difference V** across the capacitor is as large as the **source's e.m.f. E**. The capacitor is fully charged. The flow of electrons has **reduced to zero**.

The cell has done **work** to charge the capacitor. This work becomes **energy stored** in the electric field between the plates. (The field pulls on the charges.)

Work done = charge × potential difference
but the potential difference has **varied** (increased) while charging.

For a capacitor:

Energy stored = area under a Q / V graph $E = \frac{1}{2}QV$

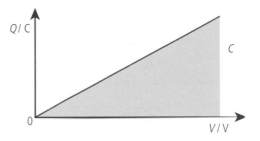

$E = \frac{1}{2}QV = \frac{1}{2}(CV)V = \frac{1}{2}CV^2$ $E = \frac{1}{2}CV^2$

$E = \frac{1}{2}QV = \frac{1}{2}Q\left(\frac{Q}{C}\right) = \frac{1}{2}\frac{Q^2}{C}$ $E = \frac{1}{2}\frac{Q^2}{C}$

1   How much charge is stored in a 5000 μF capacitor when the p.d. across it is **a)** 1 V **b)** 5 V?

   a) $Q = CV = 0.005 \times 1 = 0.005\,C$ **b)** $Q = CV = 0.005 \times 5 = 0.025\,C$

2   A capacitor takes 880 μC to charge when connected to a 4 V source. What size is the capacitor and how much energy has it stored?

   $C = \frac{Q}{V} = \frac{880 \times 10^{-6}}{4} = 220\,\mu F$ $E = \frac{1}{2}QV = \frac{1}{2}880 \times 10^{-6} \times 4 = 1760\,\mu J$

3   How much energy is stored in a 2000 μF capacitor when the p.d. across it is 4 V?

   $E = \frac{1}{2}CV^2 = \frac{1}{2}2000 \times 10^{-6} \times 4^2 = 1.6 \times 10^{-2}\,J$

4   A 1000 μF capacitor stores 400 mJ of energy. What charge has been stored?

   $E = \frac{1}{2}\frac{Q^2}{C}$ $400 \times 10^{-3} = \frac{1}{2}\frac{Q^2}{21\,000 \times 10^{-6}}$ $Q = 0.028\,C$

LET'S THINK ABOUT THIS

1   An insulating material placed between the plates of a parallel plate capacitor is called a **dielectric**. It increases the capacitance and also allows metal sheets to be rolled without shorting.

2   How much foam can you fill a beaker with? It depends on the size of the beaker and how hard you push. Like a capacitor with charge?

3   A capacitor is like a source without internal resistance, as the charges never cross the capacitor. It can release its energy over a very short time.

CAPACITORS IN D.C.

CHARGING AND DISCHARGING A CAPACITOR

Current and **voltage graphs** can be obtained while **charging** and **discharging** a **capacitor** in a **R-C circuit** by connecting the circuit to a computer and interface. The **voltages** are sampled, and the software (using Ohm's Law) calculates the **current**. The value of the resistor used has to be entered. Alternatively, you can obtain readings using voltmeters, ammeter and a stop-clock. Note that meters are not required when the computer is used.

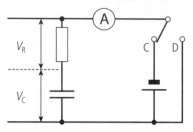

At C: the capacitor is **charging** through R.

At D: the capacitor is **discharging** across R.

The capacitor is **charged** then **discharged**.

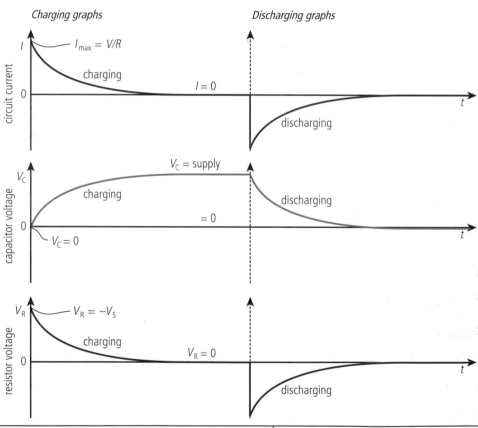

Charging graphs   *Discharging graphs*

circuit current: $I_{max} = V/R$, charging, $I = 0$, discharging

capacitor voltage: $V_C = $ supply, charging, $V_C = 0$, $= 0$, discharging

resistor voltage: $V_R = -V_S$, charging, $V_R = 0$, discharging

Charging	Discharging
• The **voltage** is across the **resistor** at the **start**. • The **voltage** is across the **capacitor** when **charged**. • These voltages always **add** to equal the size of the **source voltage**: $-V_S = V_C + V_R$ • The **current** is **zero** after $V_C = $ supply voltage.	• The capacitor **voltage** is in the **same direction** as when charging. • The **current** is in the **opposite** direction as when charging.

Try searching *capacitance* at www.wikipedia.org.

DON'T FORGET

When drawing graphs, add the values for V_{max} and I_{max}.

$v = iR$ The **current** varies as the **voltage** across the **resistor** at all points in **time**.

CHARGING TIME

Increasing the **capacitance** **increases** the charging **time**.

(More charge will be stored. $Q = CV$)

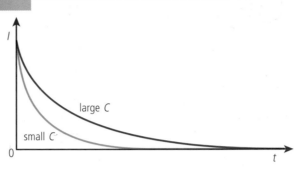

Increasing the **resistance** **increases** the charging **time**.

(The same charge will be stored.)

The current starts smaller but goes on for longer.

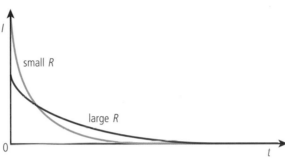

The **time** to **charge or discharge** depends on the **resistance** and the **capacitance**.

(The product of **R** × **C** is useful when comparing charging times.)

DON'T FORGET

$I_{max} = \dfrac{V}{R}$

SOLVE IT

Find:

1 the initial current at switch-on.

2 the voltage of the charged capacitor.

3 the voltage across the resistor when the voltage across the capacitor is 12 V.

4 the charge stored on the capacitor.

5 the effect of:
   • decreasing R   • increasing C   • increasing V_s

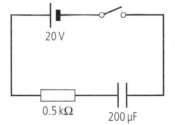

20 V

0.5 kΩ   200 μF

Answers

1 $I_{max} = \dfrac{V}{R} = \dfrac{20}{500} = 0.04\,A$

2 $V_C = V_S = 20V$

3 $V_R = 20 - 12 = 8V$

4 $Q = CV = 200\mu \times 20 = 4000\,\mu C$

5 charging time decreases, increases, stays the same

LET'S THINK ABOUT THIS

1 The charge/discharge graphs show **exponential** growth or decay.

2 If the resistor is replaced with a wire when discharging, the **time** will be **shorter** and the **maximum current** will start **larger**.

3 The circuit with the greatest **product** of **R** × **C** will take the longest time to charge.

CAPACITORS:
A.C. AND APPLICATIONS

FREQUENCY AND CURRENT

In an **a.c. circuit**, the **supply voltage** is constantly **altering direction**. This means a capacitor will always be **charging** one way, then **discharging** and **charging** in the opposite direction.

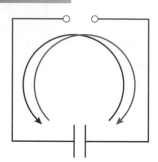

This constant **charging**, **discharging** and **recharging** means there is always a **current** in the **circuit** even though no charge crosses the capacitor.

Investigating frequency and current.

Using a **signal generator** as the power supply allows the **frequency** of the a.c. signal to be **altered**. The **frequency** can be recorded from the signal generator frequency **dial**. An **a.c. voltmeter** can be used across the supply to ensure that the **voltage does not change**, and an **a.c. ammeter** is used to take readings of **current**. An alternative to the voltmeter is to use a **cathode ray oscilloscope** to check the **voltage** and calculate the **frequency** supplied.

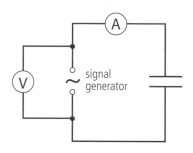

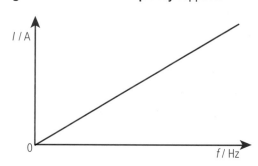

As the frequency of the supply is increased, the values of r.m.s. current also increase.

The graph shows a straight line through the origin.

When the p.d. across the capacitor is kept constant,

the **current** is **directly proportional** to the **frequency**: $\boxed{I \propto f}$

It could be said that a capacitor **opposes** current at **low frequencies** though it does not block it altogether. The **opposition decreases** as **frequency increases**, and so **current increases**.

(An extra investigation would be to find how the size of capacitor affects the a.c. current. You would find $I \propto C$. A bigger capacitor allows more current in the circuit.)

A.C./D.C. SUMMARY

In a **d.c. circuit**, once a capacitor has been fully charged, **no** more **current** exists.

A capacitor **blocks d.c.**

In an **a.c. circuit**, a current exists as the capacitor is constantly charging and discharging. This current is **proportional** to the frequency. The capacitor can be said to **pass** high frequencies.

A capacitor **passes high frequencies of a.c.**

A capacitor **blocks low frequencies and d.c.**

DON'T FORGET

Memorise this summary: it's the key to capacitance.

CAPACITOR APPLICATIONS

Applications of a **capacitor** make use of a capacitor's ability to **store energy, store charge** and **block d.c.** while **passing a.c.**

The strobe lamp

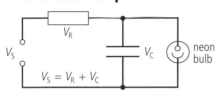

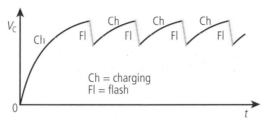

Ch = charging
Fl = flash

The strobe-lamp circuit uses a **capacitor** to **store charge** and **energy** for **sudden release** across a neon bulb. The capacitor charges through a resistor. Initially the voltage is across the resistor, but the capacitor's share increases until it is large enough for the neon bulb (which needs a high voltage) to conduct. After the neon has flashed, the process can repeat.

Filters for blocking d.c. and passing a.c.

In some circuits (such as in a radio), you find that a d.c. signal has been added to an a.c. signal. The capacitor blocks the d.c. components and passes the a.c. component.

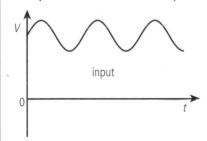

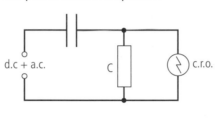

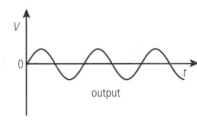

Filters for crossover networks in loudspeakers:

The capacitor **passes high frequencies** but **blocks low frequencies** to prevent damage to the 'tweeter'.

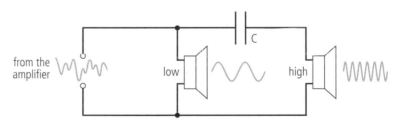

Smoothing rectified voltage

Half-wave rectification produces d.c. in pulses.

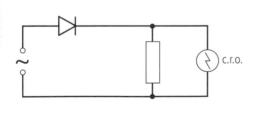

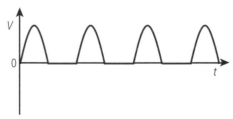

A smoothing capacitor **stores charge** during the pulse, **when the diode conducts**, then **supplies the charge** during the gaps, **when the diode does not conduct**.

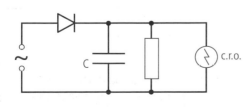

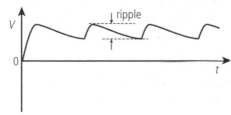

The pulses have been smoothed to a 'ripple'.

DON'T FORGET

a.c. only occurs when the voltage crosses the x-axis.

ELECTRICITY AND ELECTRONICS

OPERATIONAL AMPLIFIERS AND INVERTING MODE CIRCUITS

OPERATIONAL AMPLIFIERS

Operational amplifiers (known as **op-amps**) are **integrated circuits** designed to **amplify voltage**. The **output current** is **very small**, and these circuits are used to **amplify voltages** in **measuring instruments** or many other **electronic circuits** rather than to power a loudspeaker.

The basic op-amp characteristics are:

- **Low power**
- **Cheap** and **reliable**
- **d.c.** or **a.c.** inputs
- Extremely **high gain** (typically over 200 000)
- **2 inputs** (called **inverting** (−) and **non-inverting** (+))
- **1 output**
- dual-rail power supply (e.g. +15V, 0V, −15V)

An **op-amp** can be used to **increase** the **voltage** of a **signal**.

The **gain** of an op-amp on its own is so **high** that the output does not have a lot of use. Op-amps are usually constructed into **circuits** with a **feedback resistor** to **control the gain**.

DON'T FORGET

The supply pins are often omitted from circuit diagrams.

THE IDEAL OP-AMP

For the ideal op-amp:

- the **input current** is **zero**, i.e. it has **infinite input resistance**.

- there is **no potential difference** between the inverting and non-inverting inputs, i.e. both **input pins** are at the **same potential**.

If the resistance between the inputs is infinite and there is virtually no p.d. across the inputs, the gain is controlled by the external resistors of the circuits.

THE INVERTING MODE CIRCUIT

The **inverting mode circuit** has the **output** connected to the **inverting input** with a **feedback resistor** in a **feedback loop**. The **non-inverting input** is connected to **zero** (or 'earth'), so both input pins will be at 0 V.

The **inverting mode circuit** provides an output whose **gain** is set by the **ratio** of the feedback and input resistors:

$$\text{gain} = \frac{V_0}{V_1} = -\frac{R_f}{R_1}$$

or

$$V_0 = -\frac{R_f}{R_1}V_1$$

An **op-amp** connected in **inverting mode** will **invert** the **signal**.

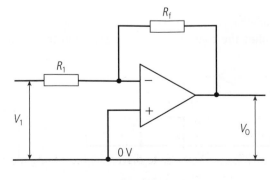

INVERTING MODE CIRCUITS

An **op-amp** is connected as shown.

Note that the **non-inverting input** is connected to **earth**.

What mode is the op-amp being used in?

This is an **inverting mode** circuit.

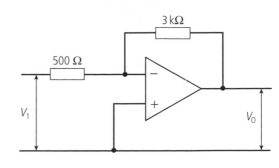

What is the output voltage if the input is 400 mV?

$$V_0 = -\frac{R_f}{R_1}V_1 = -\frac{3000}{500}\,0.400 = -2.4\,V$$

If an a.c. signal is input, you can sketch an output graph by calculating the peak output value. Remember to check you have an inverted graph.

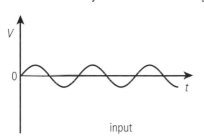
input

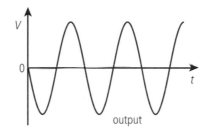
output

> **DON'T FORGET**
> The input is a potential difference between the input and earth line.

SATURATION

An amplifier uses its power supply to provide a larger copy of the input signal. Thus the **maximum output voltage** cannot exceed the supply voltage.

An **op-amp cannot** produce an **output voltage greater** than the **positive supply voltage** or **less** than the **negative supply voltage**.

The graph shows how the **d.c. output** is **limited**.

The y-axis will have a larger voltage scale than the x-axis, dependent on the gain.

In practice, the **maximum output voltage** is often only about 85% of the supply.

The a.c. output may only be limited at its peak.

This is called clipping or distortion.

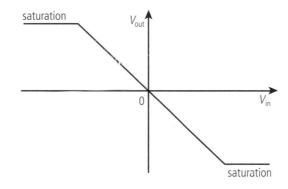

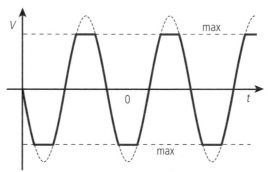

> **DON'T FORGET**
> The **op-amp saturates**. You will get **no marks** if you say the voltage saturates!

LET'S THINK ABOUT THIS

An amplifier can be made to **distort** by **increasing** the **input signal** or **increasing** the **gain**.

A **large** a.c. input with a **high-gain** op-amp may produce a **square wave**.

DIFFERENTIAL MODE CIRCUITS

THE DIFFERENTIAL MODE CIRCUIT

The **differential mode circuit** uses both the **inverting** and **non-inverting inputs** to the op-amp.

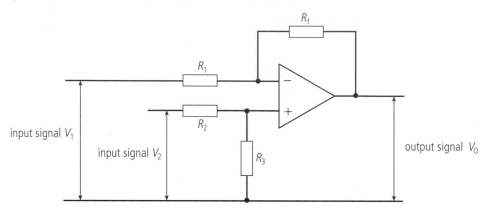

The ratio $R_f : R_1$ is the gain of V_1 while the ratio of $R_3 : R_2$ is the gain of V_2.

The **ratio** of $R_f : R_1$ and the ratio of $R_3 : R_2$ will be kept the **same** i.e. $\dfrac{R_3}{R_2} = \dfrac{R_f}{R_1}$.

The **output voltage** will be given by: $\quad V_0 = (V_2 - V_1)\dfrac{R_f}{R_1}$

The **differential amplifier** amplifies the **potential difference** between its **two inputs**.

V_1 is being applied to the **inverting input**, and V_2 is being applied to the **non-inverting input**. In the equation, V_1 has the negative in front of it, and we must remember it is V_1 that is subtracted from V_2.

You can check the voltages have been entered correctly if you remember that:

- if the **inverting (–) input** is **larger**, the output is **negative (–)**
- if the **non-inverting (+) input** is **larger**, the output is **positive (+)**.

Note that, if all the **resistors** in this circuit are the **same**, the circuit will perform **subtraction**.

In general, this circuit is performing a mixture of **multiplication** and **subtraction**.

DIFFERENTIAL OR INVERTING MODE

When approaching an op-amp question, the first thing you should check is whether the circuit is operating in **differential mode** or in **inverting mode**. To do this, you need to check whether **both inputs are being used** or whether the **non-inverting input is connected to earth**. Both circuits may be used in the same question!

Differential mode:

$$V_0 = (V_2 - V_1)\dfrac{R_f}{R_1}$$

Inverting mode:

$$gain = \dfrac{V_0}{V_1} = -\dfrac{R_f}{R_1} \quad \text{or} \quad V_0 = -\dfrac{R_f}{R_1}V_1$$

DON'T FORGET

The inputs to the op-amps can be either positive or negative.

DIFFERENTIAL MODE CIRCUITS

An **op-amp** is connected as shown.

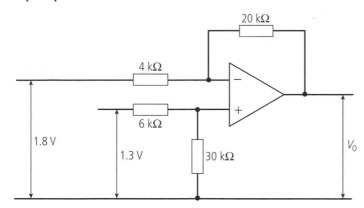

Note that both the **non-inverting input** and the **inverting input** are **connected** to resistors.

What mode is the op-amp being used in?

This is a **differential mode** circuit.

What is the **output voltage** for the inputs shown?

$$V_0 = (V_2 - V_1)\frac{R_f}{R_1} = (1.3 - 1.8)\frac{20\,k}{4\,k} = -2.5\,V$$

In this same circuit, what would the output voltage be if the input voltages were **changed** to:

$V_1 = 2.5V,$

$V_2 = 0.3V$ and the feedback resistor was changed to $100\,k\Omega$ and the $30\,k\Omega$ is changed to $150\,k\Omega$?

$$V_0 = (V_2 - V_1)\frac{R_f}{R_1} = (0.3 - 2.5)\frac{100\,k}{4\,k} = -55\,V$$

However, the **op-amp cannot** produce an **output voltage less** than the **negative supply voltage**. If the negative supply voltage is $-15\,V$, then the op-amp must be **saturated**.

In practice, this op-amp will output a voltage of around $-13.5\,V$. This is 85% of the negative supply voltage.

What assumption has been checked before using the equations above?

The **ratios** of $R_f : R_1$ and the ratio of $R_3 : R_2$ are the **same**.

What is the **gain** of this op-amp circuit?

$$gain = \frac{R_f}{R_1} = \frac{20\,k}{4\,k} = 5 \qquad \text{Note that there are } \textbf{no units} \text{ for gain.}$$

DON'T FORGET

There are only **two main equations** you need to use in op-amps. Make sure you learn them!

LET'S THINK ABOUT THIS

Operational amplifiers are one of the most common circuits in electronics.

Some slightly more advanced circuits to look out for would be:

- summing amplifier
- square wave generator
- digital timing pulse generator
- digital-to-analogue converter (DAC).

MONITORING AND CONTROL CIRCUITS

MONITORING CIRCUITS

The **differential amplifier** is very good at **amplifying small differences** in **voltage**. In the **Wheatstone bridge** (see p. 46), an **out-of-balance voltage** was produced across the voltmeter proportional to a change in resistance in a sensor. The sensor detects **physical quantities** such as **strain** or **light** or **heat**.

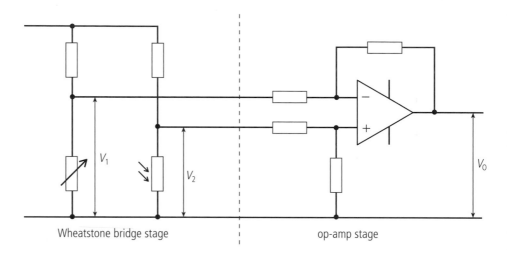

Wheatstone bridge stage   op-amp stage

DON'T FORGET

The device will still have to be **calibrated**. The output voltage will have to be **graphed** against the light or other physical quantity once, then the light level can be given for any voltage output.

DON'T FORGET

The monitoring circuit is just a Wheatstone bridge plus op-amp in differential mode.

The **op-amp**, used in a **differential mode circuit**, can replace the voltmeter in the Wheatstone bridge. The output from the **Wheatstone bridge** becomes the input to the **op-amp**. The **small out-of-balance voltage** produced by the **Wheatstone bridge** will be **amplified** by the **op-amp circuit**.

Initially, the **variable resistor** is adjusted to **balance the bridge** ($V_1 = V_2$, output $= 0V$) at a **desired level** of light.

As the light changes, the resistance of the LDR changes, and V_1 changes.

For example, as light increases, R_{LDR} decreases, and V_1 decreases. Therefore $V_2 - V_1$ is not zero.

The **voltage difference** is now **amplified** by the **op-amp**.

The op-amp has made it easier to **monitor** the voltage difference produced by the change in light.

An **LDR** has been shown operating with light, but this could be replaced with **any sensor** whose **resistance changes** (see p. 47).

What else could be monitored?

* blood pressure

* alarm system.

CONTROL CIRCUITS

An **op-amp** is a **voltage amplifier**; it cannot give out a high current.

A **transistor** is a **current amplifier**.

A transistor only switches on above a certain voltage, typically 0.7 V for an npn transistor.

An op-amp can be combined with a transistor to **control** devices (such as switching heaters or lights on or off).

The output voltage from an op-amp can be used to **switch** on the transistor, and a **larger current** will be drawn from the **transistor's power source**.

The Wheatstone bridge can also be kept in, so we can also **combine monitoring circuits** with **control circuits**.

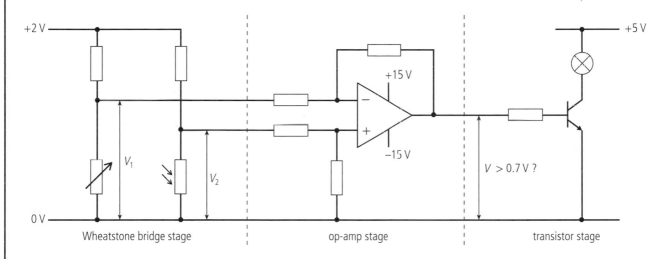

| Wheatstone bridge stage | op-amp stage | transistor stage |

This circuit will produce a small voltage difference at the **Wheatstone bridge** when the light changes. The **op-amp** amplifies this to produce a larger voltage output, which switches on the **transistor**, and the lamp will conduct.

The lamp in this circuit could be replaced by a **relay**. A relay would allow even higher power circuits to operate.

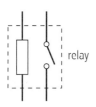

relay

> **DON'T FORGET**
>
> In this circuit, the op-amp is always in differential mode.

LET'S THINK ABOUT THIS

1 A pnp transistor would have to be more negative than –0.7 V to conduct.

2 If a motor is driven in one direction by an npn transistor, it will be driven in the reverse direction by a pnp transistor.

3 As well as npn transistors, it would be possible to use MOSFET transistors, see p. 83.

KEY QUESTIONS

REVISION QUESTIONS

Electric fields and resistors in circuits

1 What does an electric charge experience in an electric field?
2 What happens when an electric field is applied to a conductor?
3 What has to be done to move charge in an electric field?
4 Define potential difference.
5 Define the volt.
6 If a battery supplies 18 J of energy to 2 C of charge, what is its potential difference?
7 Define the e.m.f. of a source.
8 State what an electrical source is equivalent to.
9 Describe the principles of a method of measuring the e.m.f. and the internal resistance of a source.
10 Explain why the e.m.f. of a source is equal to the open circuit p.d. across the terminals of the source.
11 Explain how conservation of energy leads to the sum of the e.m.f.s round a closed circuit being equal to the sum of the p.d.s round the circuit.
12 Derive the expression for the total resistance of any number of resistors in series, by consideration of the conservation of energy.
13 Derive the expression for the total resistance of any number of resistors in parallel by consideration of the conservation of charge.
14 In a Wheatstone bridge circuit, if R1 $= 2\,\Omega$, R2 $= 4\,\Omega$ and R3 $= 6\,\Omega$, what is the value of R4?
15 What happens for an initially balanced Wheatstone bridge, as the value of one resistor is changed by a small amount?
16 Remember to carry out many calculations involving potential differences, currents and resistances in circuits containing resistors!

Alternating current and voltage

17 Describe how to measure frequency using an oscilloscope.
18 If a power pack indicates 7 Vrms, what is the peak voltage?
19 If the peak value of an ac current is 5 A, what is the r.m.s. current value?
20 State the relationship between current and frequency in a resistive circuit.

Capacitance

21 State the relationship between charge and p.d. between two parallel conducting plates.
22 Describe the principles of a method to show the relationship above.
23 Define capacitance.
24 Define the farad.
25 If 3 V puts 21 000 µC of charge onto a capacitor, what is its capacitance?
26 Explain why work must be done to charge a capacitor.
27 How can you find the work done to charge a capacitor from a Q–V graph?
28 Give three formula for the energy stored on a capacitor.
29 A 440 pF capacitor is used with a 10 V supply. How much charge and energy are stored on the capacitor?
30 Draw qualitative graphs of current against time and of p.d. against time for the charge and discharge of a capacitor in a d.c. circuit containing a resistor and capacitor in series.
31 A 10 000 µF capacitor is charged through a 500 Ω resistor by a 9 V supply. Find the maximum current in the circuit and the maximum voltage across the capacitor. When does each occur?
32 State the relationship between current and frequency in a capacitive circuit.
33 Describe the principles of a method to show how the current varies with frequency in a capacitive circuit.
34 Describe and explain the possible functions of a capacitor: storing energy, storing charge, blocking d.c. while passing a.c.

contd

REVISION QUESTIONS contd

Analogue electronics

35 What can an op-amp be used for?

36 What are the ideal conditions for an op-amp?

37 Draw an op-amp circuit showing inverting mode.

38 What will an inverting mode circuit do to the input signal?

39 If $R_f = 100\,k\Omega$, $R_1 = 20\,k\Omega$, $V_1 = 1\,V$, what will be the output voltage from an inverting mode op-amp circuit if the power supply is 9V?

40 Define the maximum output from an op-amp.

41 Draw an op-amp being used in the differential mode.

42 What does a differential amplifier do?

43 What is the equation for an op-amp in differential mode?

44 Describe how to use the differential amplifier with resistive sensors connected in a Wheatstone bridge arrangement.

45 Describe how an op-amp can be used to control external devices via a transistor.

ANSWERS

1 In an electric field, an electric charge experiences a force.

2 An electric field applied to a conductor causes the free electric charges in it to move.

3 Work W is done when a charge Q is moved in an electric field.

4 The potential difference between two points is a measure of the work done in moving one coulomb of charge between the two points.

5 If one joule of work is done moving one coulomb of charge between two points, the potential difference between the two points is one volt.

6 9V.

7 The e.m.f. of a source is the electrical potential energy supplied to each coulomb of charge which passes through the source.

8 An electrical source is equivalent to a source of e.m.f. with a resistor in series; the internal resistance.

9 See p 45. **10** See p 45. **11** See p 43. **12** See p 43. **13** See p 43. **14** 12Ω.

15 For an initially balanced Wheatstone bridge, as the value of one resistor is changed by a small amount, the out-of-balance p.d. is directly proportional to the change in resistance.

16 Practice! **17** $f = \dfrac{1}{T}$ ... see p 48. **18** 10V. **19** 3.5A.

20 The current is unaffected by frequency in a resistive circuit.

21 The charge on two parallel conducting plates is directly proportional to the p.d. between the plates.

22 See p 50. **23** Capacitance is the ratio of charge to p.d.

24 The unit of capacitance is the farad and that one farad is one coulomb per volt.

25 7000μF. **26** See p 51.

27 The work done to charge a capacitor is given by the area under the graph of charge against p.d.

28 $E = \frac{1}{2}QV$ $E = \frac{1}{2}CV^2$ $E = \frac{1}{2}\dfrac{Q^2}{C}$ **29** $Q = 4400pC$. $E = 22\,000\,pJ$.

30 See p 52. **31** $I_{max} = 0.018A$ initially. $V = 9V$ finally.

32 The current is directly proportional to the frequency in a capacitive circuit.

33 See p 54. **34** See p 55. **35** An op-amp can be used to increase the voltage of a signal.

36 For the ideal op-amp: **a)** input current is zero, ie it has infinite input resistance; **b)** there is no potential difference between the inverting and non- inverting inputs; ie both input pins are at the same potential.

37 See pp 56–57.

38 An op-amp connected in the inverting mode will invert the input signal (plus amplify it).

39 −5V.

40 An op-amp cannot produce an output voltage greater than the positive supply voltage or less than the negative supply voltage.

41 See pp 58–59.

42 A differential amplifier amplifies the potential difference between its two inputs.

43 $V_0 = (V_2-V_1)\dfrac{R_f}{R_1}$

44 See p 60. **45** See p 61.

WAVE PROPERTIES

WAVES

Waves are created by **vibrating sources**.
A wave is a **movement** of **energy**.

Wavelength λ is the distance in which a wave repeats.

Amplitude *a* is the height to a crest or trough from rest.

The **energy** (E_P and E_K) of a wave depends on its **amplitude**.

The **period** *T* is the time for one wave. $\boxed{T = \dfrac{1}{f}}$

The **frequency** *f* is the number of waves per unit time. $\boxed{\text{frequency} = \dfrac{\text{number}}{\text{time}} \quad \text{and} \quad f = \dfrac{1}{T}}$

The **frequency** of a wave is the same as the frequency of the source producing it.

The **wave speed** *v* is the distance travelled in unit time. $\boxed{v = \dfrac{s}{t}}$

The **wave equation** relates wave speed, frequency and wavelength. $\boxed{v = f\lambda}$

Reflection, **diffraction**, **refraction** and **interference** are characteristic behaviours of all types of **waves**.

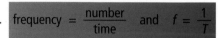

REFLECTION

$\boxed{i = r}$

Note: *v*, *f* and λ are unchanged.

Although the light rays **reflect**, they appear to come from behind the mirror!

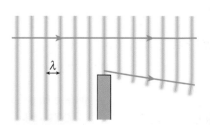

DIFFRACTION

Bending called **diffraction** occurs at **edges** or **gaps**.

Edges

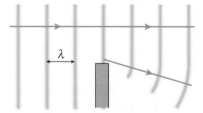

Low-frequency long wavelengths **diffract most**.

High-frequency short wavelengths **diffract least**.

Gaps

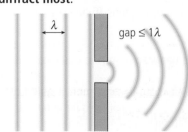

gap ≤ 1λ

gap > 1λ

Circular wavefronts are produced if the gap is ≤ 1λ.

DON'T FORGET

Energy moves through a medium, the particles just vibrate.

REFRACTION

Waves **change direction** when passing from one medium to another because the waves **change velocity**.

v changes, λ changes, frequency **f stays the same**.

$$f = \frac{v_1}{\lambda_1} = \frac{v_2}{\lambda_2}$$

If the **frequency** of a light ray changed, the **colour** would change.

Note: this ray could be travelling in either direction.

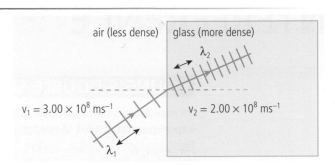

air (less dense)   glass (more dense)

$v_1 = 3.00 \times 10^8 \text{ ms}^{-1}$ $v_2 = 2.00 \times 10^8 \text{ ms}^{-1}$

DISPERSION

When **white** light **refracts**, it **disperses** into a **continuous spectrum** of its colours (frequencies).

Violet light refracts more than red light.
Infrared and ultraviolet are invisible.

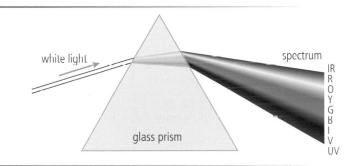

white light   spectrum

glass prism

IR
R
O
Y
G
B
I
V
UV

INTERFERENCE

When **waves meet**, they **pass through** each other. Their **energies** add up **constructively** or **destructively**. After meeting, they continue in the **same direction** as before.

Constructive interference

Waves meet **in phase**.

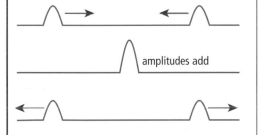

amplitudes add

Only **waves** show **interference**; particles reflect.

Destructive interference

Waves meet **out of phase**.

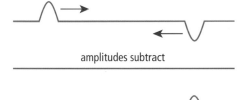

amplitudes subtract

> **DON'T FORGET**
>
> **Interference** is considered the test for a wave.

LET'S THINK ABOUT THIS

1 These wave properties apply to all types of waves.

2 How many types of wave do you know?
   Water waves, light, all the electromagnetic spectrum, sound.

3 Revise the order of the electromagnetic spectrum.

4 Learn these basic wave properties, as we'll be studying some in more detail next!

5 The frequency of middle C is 256 Hz. What is the period of its note?

$$T = \frac{1}{f} = \frac{1}{256} = 0.0039\, s$$

INTERFERENCE

YOUNG'S DOUBLE-SLIT EXPERIMENT

Thomas Young's famous **double-slit experiment** from 1801 showed that **light energy** travelled as a **wave motion**. On the screen, he saw an **interference pattern**, a **series of light and dark fringes**.

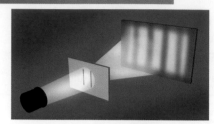

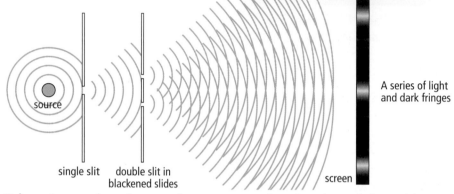

single slit   double slit in blackened slides   screen

A series of light and dark fringes

Coherent waves have the **same frequency, wavelength** and **speed** along with a **constant phase relationship**.

Using a **single light source** with the **double slits** produces **coherent waves**. **Diffraction** at each slit produces **circular wavefronts**, which spread out so that they meet and produce **interference**.

PATH DIFFERENCE

Constructive interference takes place when the two light waves are **in phase**. The waves are **in phase** when the **path difference** Δ is an **integer** number of **wavelengths**.

Constructive interference for a **maxima** or bright fringe occurs when $\boxed{\Delta = n\lambda}$
i.e. $\Delta = 0, 1\lambda, 2\lambda, 3\lambda$

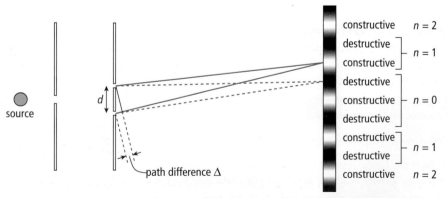

source

d

path difference Δ

constructive $n = 2$
destructive
constructive $n = 1$
destructive
constructive $n = 0$
destructive
constructive
destructive $n = 1$
constructive $n = 2$

Destructive interference takes place when the two light waves are **out of phase**. The waves are out of phase when a **crest** meets a **trough**.

Destructive interference for a **minima** or dark fringe occurs when $\boxed{\Delta = (n + \tfrac{1}{2})\lambda}$

i.e. $\Delta = \tfrac{1}{2}\lambda, 1\tfrac{1}{2}\lambda, 2\tfrac{1}{2}\lambda$

DON'T FORGET

Watch out: many students enter the wrong integer n.

MEASURING THE WAVELENGTH OF LIGHT

For a maxima, the path difference $= n\lambda$

Using trigonometry, $n\lambda = d\sin\theta$

For a minima, the path difference $= (n + \frac{1}{2})\lambda$

Using trigonometry: $(n + \frac{1}{2})\lambda = d\sin\theta$

The equation $n\lambda = d\sin\theta$ can be used to calculate the wavelengths of light. The slit separation d and the angle θ to the nth-order maxima first needs to be measured.

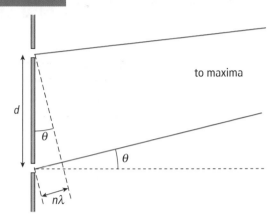

to maxima

THE RANGE OF VISIBLE LIGHT

The **wavelength** of **visible light** ranges from about 4×10^{-7} **m** to about 7×10^{-7} **m**.

Here are some wavelengths given in nanometres:

blue   480 nm   green   509 nm   red   644 nm

DON'T FORGET

You can often find wavelengths on the exam data sheet.

 Try entering *Young's Interference* into an internet search engine.

LET'S THINK ABOUT THIS

1 You can find interference with many kinds of waves.

2 **Water waves** on the ripple tank show interference. Note the lines of **rough (constructive interference)** and the lines of **calm (destructive interference)**.

3 **Two loudspeakers** connected to the one signal generator so they are **in phase** produce areas of **louder** and **quieter sound**. Two speakers producing **less** sound than one … in places!

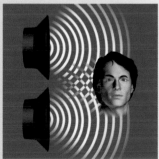

4 You have to watch out for reflections of walls. You can have **interference** between sound **direct** from one speaker and its **reflection** from a side wall.

5 Equations apply to all wave interference:

$n\lambda = d\sin\theta$ $(n + \frac{1}{2})\lambda - d\sin\theta$

DIFFRACTION GRATINGS

THE DIFFRACTION GRATING

A **diffraction grating** may look like a small piece of glass. A diffraction grating consists of a small piece of glass into which many **thousands** of **tiny parallel lines** have been etched.

After light goes through the parallel lines, the **interference pattern** appears in a similar place to that obtained from a double-slit slide. However, the peak intensities are higher and the maxima are sharp and narrow, creating a **better resolution**.

Calculating *d*

You are often given the number of **lines per millimetre** and have to work out the **spacing *d***.

Example

A grating has 500 lines per millimetre. Calculate the spacing *d* in metres.

$$d = \frac{distance}{number} = \frac{1 \times 10^{-3}}{500} = 2 \times 10^{-6}\,\text{m}$$

INTERFERENCE USING A GRATING

A **diffraction grating** and a light source can be set up to produce an **interference pattern**.

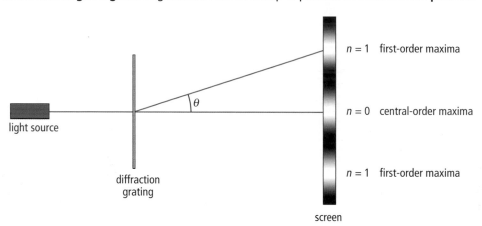

light source

diffraction grating

$n = 1$ first-order maxima
$n = 0$ central-order maxima
$n = 1$ first-order maxima

screen

In a **diffraction grating**, the **slits** are **very close** together. The spacing of the **fringes** on the screen is therefore much **greater** than with the standard double slit.

The **equations** for wave interference with a **diffraction grating** are the same as for Young's slits:

Constructive: $n\lambda = d\sin\theta$ Destructive: $(n + \frac{1}{2})\lambda = d\sin\theta$

The laser

The **laser** is a good light source to use with a diffraction grating. The light from a laser is **monochromatic** (one colour, one frequency) and is **coherent** (same v, f, λ, and all the light has the same phase).

INTERFERENCE OF WHITE LIGHT

When a **white light source** is used with a **diffraction grating**, a **series of spectra** are produced on the screen by **interference**.

A bright **central white maxima** is produced as all the wavelengths of white light interfere **constructively** at the centre.

Red light is **deviated more** than **blue light** since it has the **longest wavelength**. This can be shown from the grating equation

$n\lambda = d\sin\theta$

$\lambda_{red} > \lambda_{blue}$ so $\sin\theta_{red} > \sin\theta_{blue}$ and $\theta_{red} > \theta_{blue}$

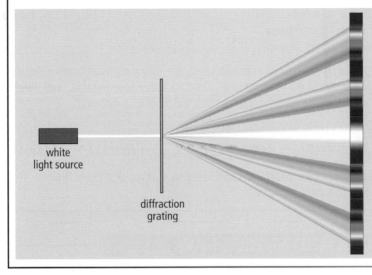

second-order spectrum

first-order spectrum

0 order (white)

first-order spectrum

second-order spectrum

white light source

diffraction grating

DON'T FORGET

Violet on the inside; red on the outside.

DISPERSION USING A PRISM

A **prism** produces a **single spectrum** by **refraction**.

Red light is deviated **least**, **violet** light is deviated the **most**.

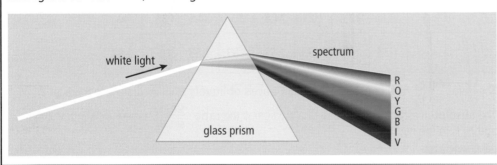

white light

spectrum

glass prism

R
O
Y
G
B
I
V

DON'T FORGET

Red is deviated the **most** in the interference experiment as shown, the **least** when refracting with the triangular prism.

LET'S THINK ABOUT THIS

1 The tracks of a CD act as a kind of reflection diffraction grating and separate out the colours of white light. There are about 625 tracks per millimetre on a CD, which is similar in spacing to the diffraction gratings used in schools.

2 A laser produces a diffraction pattern with a 100-lines-per-mm grating. If the first maxima is found at 10°, what is the wavelength and type of radiation?

$d = \dfrac{distance}{number} = \dfrac{1 \times 10^{-3}}{100} = 1 \times 10^{-5}\,\text{m}$

$n\lambda = d\sin\theta \Rightarrow 1\lambda = 1 \times 10^{-5}\sin 10° \Rightarrow \lambda = 1.74 \times 10^{-6}\,\text{m} = 1740\,\text{nm}.$

This is in the **infra-red** region.

REFRACTIVE INDEX

REFRACTION OF LIGHT

When light travels from one medium to another, it undergoes **refraction**.

When light travels from a **less dense to a more dense** material, it **slows down** and bends (or refracts) **towards the normal**. When light travels from a **more dense to a less dense** material, it **speeds up** and bends **away from the normal**. But how much does it bend?

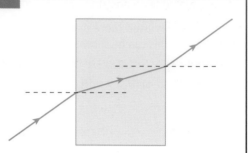

The amount of bending depends on the type of materials used. The way a material bends light is determined by the **refractive index** of the material. The refractive index of a material, with light coming from a vacuum, is known as the **absolute refractive index** of that material.

Some **typical refractive indexes** (with yellow light) are:

Diamond: 2.42   Crown glass: 1.50   Water: 1.33   Air: 1.00

When **calculating** refractive indices from **practical experiments** to three significant figures, we can use **air** instead of needing to have a **vacuum**.

REFRACTIVE INDEX AND SNELL'S LAW

When light travels from medium 1 to medium 2, study the **angles of incidence** θ_1 and **refraction** θ_2 and you find that the relationship between them is not immediately apparent. However, if you plot $\sin\theta_1$ against $\sin\theta_2$, you obtain a **straight line through the origin** showing that $\sin\theta_1$ is **directly proportional** to $\sin\theta_2$.

$\theta_1°$	0	10	20	30	40	50	60	70	80
$\theta_2°$	0	6.5	12.9	19.1	24.8	30	34.5	38	40
$\dfrac{\sin \theta_1}{\sin \theta_2}$	–	1.53	1.53	1.53	1.53	1.53	1.53	1.53	1.53

The ratio $\dfrac{\sin \theta_1}{\sin \theta_2}$ is a **constant** when light travels **obliquely** from medium 1 to medium 2.

The **absolute refractive index, n,** of a medium is the **ratio** $\dfrac{\sin \theta_1}{\sin \theta_2}$ where θ_1 is in a vacuum and θ_2 is in the medium.

$$n = \frac{\sin \theta_1}{\sin \theta_2}$$ Snell's Law

In practice, we can use: $n = \dfrac{\sin \theta_{air}}{\sin \theta_{medium}}$

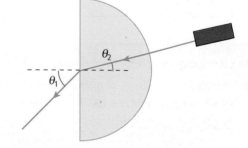

Example

A ray of light is shone on the glass block as shown. The ray emerges into air. The **angles in the medium and in the air are measured from the normal**. $\theta_2 = 25°$, $\theta_1 = 40°$.

$$n = \frac{\sin \theta_{air}}{\sin \theta_{medium}} = \frac{\sin 40}{\sin 25} = 1.52$$ The refractive index of this glass is 1.52.

REFRACTIVE INDEX AND VELOCITY

When a **wave** passes from one medium to another with a different refractive index, there is a **change in wave speed**. The **ratio of wave speeds** can be used to find the **refractive index** of a material.

$$n = \frac{v_{air}}{v_{medium}} \quad \text{or} \quad n = \frac{v_1}{v_2}$$

Examples

1. A light ray enters a glass block at 3×10^8 ms^{-1} from air and slows to 2×10^8 ms^{-1} in the glass.

$$n = \frac{v_{air}}{v_{glass}} = \frac{3 \times 10^8}{2 \times 10^8} = 1.50$$

2. If the refractive index of water is 1.33, what is the speed of light in water?

$$n = \frac{v_{air}}{v_{water}} \Rightarrow 1.33 = \frac{3 \times 10^8}{v_{water}} \Rightarrow v_{water} = 2.26 \times 10^8 \ ms^{-1}$$

REFRACTIVE INDEX AND WAVELENGTH

When a **wave** passes from one medium to another, there is a **change in wavelength**.

The **ratio of wavelengths** can be used to find the **refractive index** of a material.

$$n = \frac{\lambda_{air}}{\lambda_{medium}} \qquad n = \frac{\lambda_1}{\lambda_2}$$

$$f = \frac{v}{\lambda} = \text{constant}$$

As **v decreases**, λ **decreases**

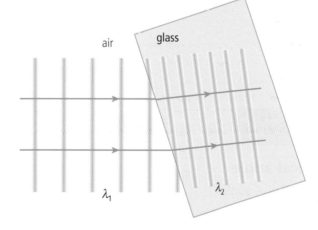

air   glass

λ_1 λ_2

DON'T FORGET

Velocity and wavelength change, frequency does not.

LET'S THINK ABOUT THIS

1. The absolute refractive index of air is 1.0003 to 5 significant figures, so it can be used as the equivalent of the refractive index in a vacuum.

2. Light paths are **reversible**.

3. Combining equations gives:
$$n = \frac{\sin \theta_1}{\sin \theta_2} = \frac{v_1}{v_2} = \frac{\lambda_1}{\lambda_2}$$

4. General equations work regardless of direction:

$$n_1 \sin \theta_1 = n_2 \sin \theta_2 \qquad n_1 v_1 = n_2 v_2 \qquad n_1 \lambda_1 = n_2 \lambda_2$$

When going from a vacuum or air to a medium, use $n_1 = 1.00$. These equations also apply between any two materials, e.g. from water to glass.

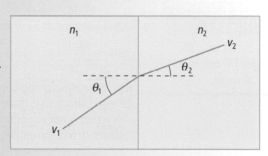

TOTAL INTERNAL REFLECTION AND COLOUR

CRITICAL ANGLE AND TOTAL INTERNAL REFLECTION

1   When a ray of light passes from an **optically dense material** to a **less optically dense** material, the **refraction** causes the ray to bend **away** from the normal. $\theta_a > \theta_m$

2   When the angle of incidence θ_m is **increased**, there will come a point where the angle of refraction will reach **90°**. The angle of incidence, which causes an **angle of refraction equal to 90°**, is called the **critical angle** θ_C.

3   With an **angle of incidence greater** than the **critical angle**, all the light energy is **reflected** and **none** is refracted. This is **total internal reflection** (TIR).

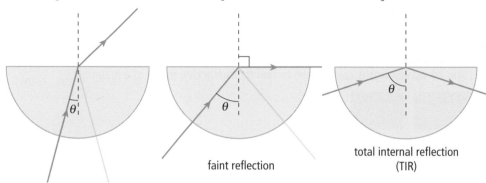

1   $\theta < \theta_C$ 2 $\theta = \theta_C$ 3 $\theta > \theta_C$

faint reflection

total internal reflection (TIR)

$$n = \frac{\sin \theta_{air}}{\sin \theta_{medium}}$$

$$\theta_{air} = 90°$$

$$i = r$$

Critical angle

At the **critical angle**,

● the angle in the medium = the critical angle, $\theta_m = \theta_C$

● the angle in air is 90°, $\theta_{air} = 90°$

● $n = \dfrac{\sin \theta_{air}}{\sin \theta_{medium}} = \dfrac{\sin 90}{\sin \theta_c} = \dfrac{1}{\sin \theta_c}$ $n = \dfrac{1}{\sin \theta_c}$ or $\sin \theta_c = \dfrac{1}{n}$

Examples

1   Calculate the critical angle for diamond, which has refractive index = 2.42.

$\sin \theta_C = \dfrac{1}{n} = \dfrac{1}{2.42} = 0.413$ Critical angle, $\theta_C = 24.4°$

2   What is the refractive index of a material whose critical angle is 48°?

$n = \dfrac{1}{\sin \theta_c} = \dfrac{1}{\sin 48} = 1.35$ Refractive index, $n = 1.35$

3   What happens to a ray incident at 55° on a glass-to-air surface? ($n_{glass} = 1.50$)?

$n = \dfrac{\sin \theta_{air}}{\sin \theta_{medium}}$ $1.50 = \dfrac{\sin \theta_{air}}{\sin 55}$ Calculator gives 'Error' for θ_{air}

This error is occurring because the equation is trying to have $\sin \theta_{air} > 1$. An impossible physical situation is being implied. The light cannot refract out greater than 90°. In fact, refraction is not occurring, and this equation no longer applies. Total internal reflection is occurring, and the angles are equal:

$\Rightarrow$ TIR occurs, $i = r$ = 55°.

DON'T FORGET

Refraction occurs at < the critical angle, TIR occurs at > the critical angle.

COLOUR

White light contains a **spectrum of frequencies**.

During **refraction**, the **higher-frequency blue rays** are **refracted more** than the **lower-frequency red rays**.

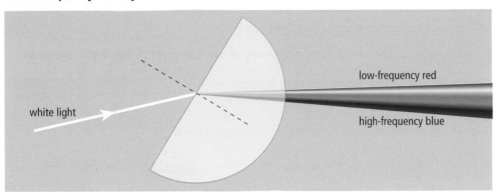

low-frequency red

white light

high-frequency blue

The **refractive index** depends on the **frequency** of the incident light.

$$n = \frac{\sin \theta_{white}}{\sin \theta_{red}} \qquad n = \frac{\sin \theta_{white}}{\sin \theta_{blue}}$$

Thus the **refractive index** of a material may **vary** depending on the **colour or frequency** of light in use. (The variation could be from 1.51 to 1.53 in glass.)

 **DON'T FORGET**

'Chromatic aberration' is due to **different refractive indices** for **different colours**.

 Try entering *Snell's Law* into an internet search engine.

LET'S THINK ABOUT THIS

1 There are many options for equations:

$$n = \frac{\sin \theta_1}{\sin \theta_2} = \frac{\sin \theta_{air}}{\sin \theta_{medium}} = \frac{v_1}{v_2} = \frac{\lambda_1}{\lambda_2} = \frac{1}{\sin \theta_C}$$

2 If **Jaws** is swimming close to the surface in front of your boat, **he can't see you**, but also **you can't see him**! Even with a torch! Until the situation gets closer than **critical**!

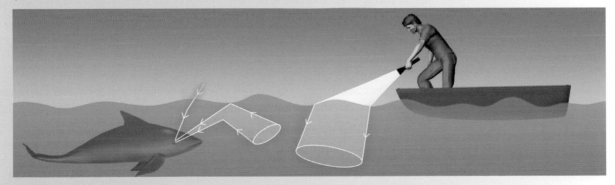

3 How would you calculate the **range of dispersion** (angle between red light and blue light) above? First calculate the angles θ_{red} and θ_{blue}, then subtract.

4 **Optical fibres** carry signals for **telecommunications**. They rely on **total internal reflection**, and the signal travels a long distance. Normally, **monochromatic light** is used. As the **different colours have a different refractive index**, this means they would travel at slightly **different speeds**. The information would not stay together and the communication would be lost.

THE PHOTOELECTRIC EFFECT

THE PHOTOELECTRIC EFFECT

Interference experiments prove that **light is a wave motion**, but here is an experiment which seems to prove something different!

When certain **metals** are exposed to **high frequencies** of electromagnetic **radiation** (such as **ultra-violet** light), **electrons** are **ejected** from the surface. This interaction between light and electrons is called the **photoelectric effect**.

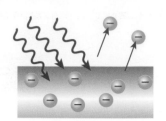

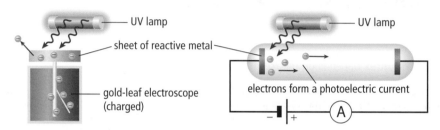

- The **metal** (e.g. clean zinc) is **negatively charged** (many electrons). Note that this effect does not work if the metal is made positive.

- Expose the metal to short-wavelength, **high-frequency** rays (**ultra-violet**). Note that this effect does not work with longer-wavelength rays even if these are very bright.

- The gold-leaf electroscope **discharges**, or there is now a **photoelectric current**.

Apparent effects

- **Electrons** are **ejected**. (Reactive metals do have outer electrons that are easy to eject.)

- There is a minimum **threshold frequency f_0** for this effect to occur.

- Light **cannot be a continuous wave** – otherwise a bright, low-frequency wave would have enough energy to eject electrons.

- The **light appears** to behave as a stream of **particles**; each UV particle has enough energy to eject an electron whereas each low-frequency particle (such as from red light) does not.

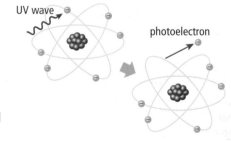

Conclusions

- **Photoelectric emission** from a surface occurs only if the **frequency** of the incident radiation is **greater** than some **threshold frequency** f_0 which depends on the nature of the surface.

- For **frequencies smaller** than the threshold value, an increase in the **irradiance** of the radiation at the surface will **not** cause **photoelectric emission**.

For classical **wave** theory, the **irradiance** should have an effect and the **frequency** should not!

WAVE–PARTICLE DUALITY

Interference experiments suggest a **wave** nature of light, but the **photoelectric effect** suggests a **particle** nature of light. Both natures exist, and this is called **wave–particle duality**.

A beam of **radiation** can be regarded as a stream of **individual energy bundles** called **photons**, each having an **energy** dependent on the **frequency** of the radiation.

> **DON'T FORGET**
>
> Light can behave as a wave or as a particle.

THRESHOLD FREQUENCY AND WORK FUNCTION

Max **Planck** was the first to show that **energy** was dependent on wavelength or **frequency**. Albert **Einstein** then had a breakthrough insight that Planck's idea made sense when light is described as **discrete light particles**, which he called **quanta** or **photons**. We can say light is **quantised**.

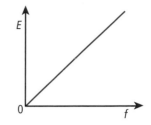

The **energy** of **each** quantum of light, or **photon**, is given from:

$E \propto f$ $\boxed{E = hf}$ where h is a constant

$f = \dfrac{c}{\lambda}$ so $\boxed{E = h\dfrac{c}{\lambda}}$ Planck's constant, $h = 6.63 \times 10^{-34}$ J s and $c = 3.00 \times 10^8$ ms^{-1}.

Example

What is the energy of a photon of blue light of wavelength 480 nm?

$$E = h\frac{c}{\lambda} = 6.63 \times 10^{-34}\,\frac{3.00 \times 10^8}{4.80 \times 10^{-7}} = 4.14 \times 10^{-19}\,J$$

Einstein understood the **work function** w_0 as the amount of **energy** the electron needs to absorb in order to be released from the atom. In the photoelectric effect, an **electron** will absorb the energy of a **single photon**. If the energy of the photon is **greater** than the **work function**, the electron will be **ejected**. The minimum frequency of a **photon**, which will have this energy, is the **threshold frequency** f_0. The **work function** is defined as $\boxed{w_0 = hf_0}$. The work function and threshold frequency are different for different metals.

A **photon** with a **frequency higher** than the **threshold frequency** will have **energy** which is **greater** than the **work function**. The **excess** energy supplies the electron with **kinetic energy**:

Energy of photon = work function + kinetic energy of electron: $\boxed{hf = hf_0 + \tfrac{1}{2}mv^2}$

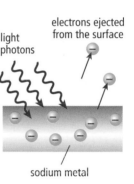
electrons ejected from the surface
light photons
sodium metal

Photoelectrons are ejected with a **maximum kinetic energy**, which is given by the difference between the **energy** of the **incident photon** and the **work function** of the **surface**.

Kinetic energy of the electron: $\boxed{E_k = hf - hf_0}$

PHOTOELECTRIC CURRENT

For **frequencies** greater than the **threshold value**, the **photoelectric current** produced by monochromatic radiation is **directly proportional** to the **irradiance** of the **radiation** at the surface.

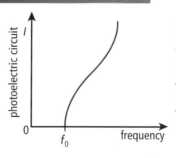
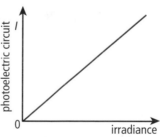

> ### DON'T FORGET
> A **bright** light source produces **more photons** per second causing **more electrons** to be **ejected**, but only if the **frequency** is **high enough**.

If **N photons per second** are incident **per unit area** on a surface, the **irradiance** at the surface is

$\boxed{I = Nhf}$ N = number of photons per second per square metre.

For more information enter *photoelectric effect* into an internet search engine.

LET'S THINK ABOUT THIS

If the ejected electrons are captured and stopped by an electric field, their **velocity** can be found from $\boxed{qV = \tfrac{1}{2}mv^2}$ (see Electrics topic). *V* is called the **stopping potential**.

ATOMIC MODELS AND LIGHT

ENERGY LEVELS

Rutherford's model of the atom was a **central positive charge** with **negative electrons** around. Niels **Bohr** used the ideas of Einstein and Planck to extend this model to suggest that <u>electrons in a free atom occupy **discrete energy levels**</u>.

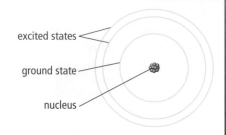

- Electrons are placed in circular orbits around the nucleus.

- There are a limited number of allowed energy levels, so the orbitals are quantised.

- An electron, bound in an atom, occupies certain states, equal to the allowed levels.

An electron may gain energy and jump to an excited state. The electron may then drop from an excited state to a lower state, emitting energy as a quantum or photon of radiation.

- An electron can be bound to the nucleus in the **ground state** or a number of **excited states**. All these energy levels for bound electrons have **negative values**.

- An electron in the ground state has the least energy. The ground-state energy level is a measure of the energy needed to **unbind** or **ionise** the electron.

- An electron just freed from the atom is then said to have reached zero energy value.

- Once an electron is free from an atom, it gains positive kinetic energy.

> **DON'T FORGET**
>
> Not all the energy levels are shown. The higher levels get closer together.

> **DON'T FORGET**
>
> The data book uses W for energy levels, e.g.
>
> $$W_2 - W_1 = hf$$

Energy levels of the hydrogen atom

- The electron has least energy in the ground state, closest to the nucleus.

- An electron moves to a higher energy level when it absorbs the energy of a photon.

- When a photon is absorbed, an electron jumps, or makes a transition, from a lower energy level to a higher energy level.

- The electron can only absorb photons of certain energies exactly matched to the energy difference between two energy levels.

- When an electron makes a transition from a higher energy level to a lower energy level, a photon is emitted.

$$E_{photon} = hf = E_{excited} - E_{lower}$$

- The frequency of the absorbed or emitted photon can be found from:

$$f = \frac{E_{excited} - E_{lower}}{h} \quad \text{or} \quad f = \frac{\Delta E}{h}$$

Note that the minus signs can be ignored if you wish, as long as you subtract the size of the two numbers to get the change in energy ΔE. It is however done properly below.

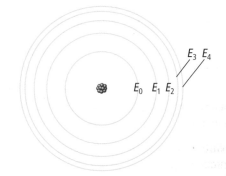

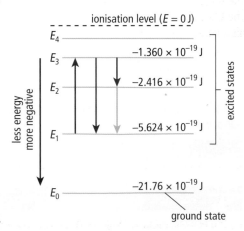

Example

A **photon** is **emitted** when an **electron** makes a **transition** from an energy level of -2.416×10^{-19} J to a lower level of -21.76×10^{-19} J. Calculate the frequency of the photon.

$$f = \frac{\Delta E}{h} = \frac{-2.416 \times 10^{-19} - (-21.76 \times 10^{-19})}{6.63 \times 10^{-34}} = 2.918 \times 10^{15} \text{ Hz} \quad \text{An ultra-violet photon.}$$

SPECTRA

The **spectrum** from a light source can be displayed using a **prism** or a **diffraction grating**.

An **emission line** in a **spectrum** will be seen when **electrons** make a **transition**, emitting photons between an **excited energy level** and a **lower energy level**.

The energy to **raise** an electron to an excited state can come from:
- a high voltage in discharge tubes
- heat in a filament lamp
- nuclear fusion in the stars.

Line-emission spectrum

Line-emission spectra give an insight into the **structure** of an **atom**. Each **element** has a **different line spectrum**, as each has a **different** structure of **energy levels**.

From four energy levels, six **different frequencies** of photons could be **emitted** and six lines of light can be seen on the **line spectrum**.

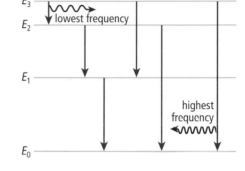

$$\Delta E = hf$$

Line-emission spectra are emitted from **low-pressure gases** where there are **free atoms**. **Lines** are **brighter** where **more electrons** make a **certain transition**.

Continuous-emission spectrum

Continuous-emission spectra are seen from **solids**, **liquids** and **high-pressure gases**. Their electrons are bonding with other atoms, giving **infinite transitions** and **infinite lines**.

Line-absorption spectrum

The **absorption spectrum** of an element has **black lines** on a **continuous spectrum**, which are in the identical position to the bright lines of that element's line emission spectrum.

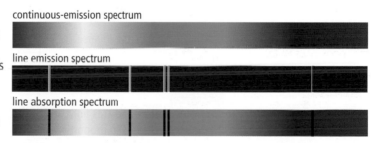

continuous-emission spectrum

line emission spectrum

line absorption spectrum

An **absorption line** in a spectrum occurs when **electrons** in a lower energy level **absorb radiation** and are excited to a higher energy level.

White light produced from a lamp or the sun has **certain frequencies missing** after passing through a gas cloud. The **gas absorbs** the frequencies of light, which it would normally emit. The gas's electrons absorb the photons whose energies match the energy transitions available. The **black lines** are known as **Fraunhofer lines** after the scientist Joseph von Fraunhofer, who discovered these lines missing from the sun's spectrum due to gases in the atmosphere.

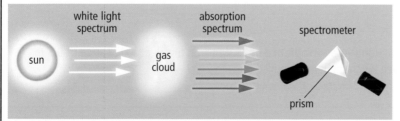

white light spectrum

absorption spectrum

spectrometer

sun

gas cloud

prism

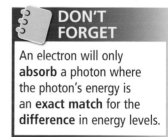

DON'T FORGET

An electron will only **absorb** a photon where the photon's energy is an **exact match** for the **difference** in energy levels.

LET'S THINK ABOUT THIS

1 If electrons make a **large energy transition**, the photon frequency may be high, so that **ultra-violet** or even **X-rays** are emitted.

2 After electrons **absorb photons** from **one direction** to create black Fraunhofer lines, they fall back down again and **emit photons**, but this time the photons can be emitted in **all directions** and so are not seen.

IRRADIANCE AND THE LASER

IRRADIANCE

Irradiance is the term used when electromagnetic **radiation** is incident on a **surface**. **Irradiance** is a measure of the **power** incident on a **surface** and is measured in Wm^{-2}.

Irradiance is the **power per unit area** $\quad I = \dfrac{P}{A}$

Example

A 150 W lamp illuminates a screen whose area is 4 m^2. Calculate the irradiance on the screen.

$$I = \frac{P}{A} = \frac{150}{4} = 37.5 \ Wm^{-2}$$

IRRADIANCE AND DISTANCE

Irradiance decreases as you move further away from a point source.

The relationship between **irradiance** and **distance** can be investigated using a **metre stick** and **photo-diode** (or light sensor) connected to an **interface** and **computer**.

Irradiance follows the Inverse Square Law:

 $\quad I \propto \dfrac{1}{d^2} \qquad I_1 d_1^2 = I_2 d_2^2$

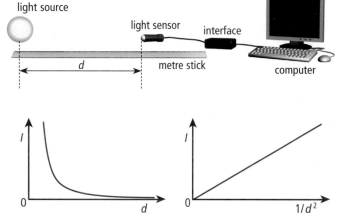

If a **point source** radiates light uniformly in **all directions** and there is no absorption, then the irradiance decreases in proportion to the **square** of **distance** from the object, since the **power** is constant and it is spread over an **area** that increases with the **square** of the **distance**.

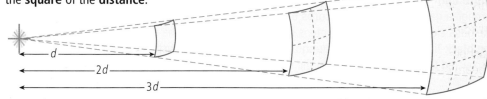

Example

A **point-source** lamp produces an **irradiance** of 25 Wm^{-2} on a wall at a **distance** of 0.5 m. What is the **irradiance** when the lamp is pulled back to 2 m?

$$I_1 d_1^2 = I_2 d_2^2 \quad 25 \times 0.5^2 = I_2 \, 2^2 \quad I_2 = 1.56 \ W \ m^{-2}.$$

STIMULATING LIGHT

Spontaneous emission

Light is **emitted** when **electrons** in an **excited state** drop to a **lower energy level**. The drop can be **spontaneous**, happening at a **random time**, and **emitting photons** in **any direction**. This is how most **ordinary sources** make light.

Spontaneous emission of radiation is a **random** process analogous to the **radioactive decay** of a **nucleus**.

contd

STIMULATING LIGHT contd

Stimulated absorption

If an **incident photon** is **absorbed**, an electron takes **all** its **energy**.

The **photon energy** is **absorbed** and the electron will jump to a **higher energy level** if the photon's **frequency** gives it an energy which is an **exact match** for the **difference** in **energy levels**. $hf = E_1 - E_0$ or $hf = W_1 - W_0$

In normal **crystals** and **gases**, there are **many electrons** in the **lower energy levels**, and it is easy for **photons** to be **absorbed**.

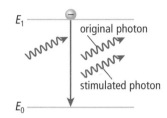

Stimulated emission

When radiation of **energy** *hf* is incident on an excited atom, the atom may be **stimulated** to emit its **excess energy** *hf*.

If an electron is in the **excited state**, and the incoming photon's energy matches the difference in energy levels, the photon will **stimulate** the electron to fall and **emit** a photon of **identical frequency**.

In **stimulated emission**, the **incident radiation** and the **emitted radiation** are **in phase** and travel in the **same direction**.

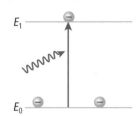

THE LASER

The conditions in the **laser** are such that a light beam **gains more energy** by **stimulated emission** than it loses by absorption – hence **L**ight **A**mplification by the **S**timulated **E**mission of **R**adiation. There must be a **population inversion** (more electrons in the **excited state**) in the laser medium. A photon starts a **chain reaction**, as emitted photons become stimulating photons.

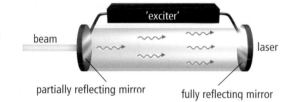

The laser has **mirrors** at each end. The photons are **reflected** back and forth creating an **avalanche** effect, and this **amplification** creates a **powerful pulse** of light. Some light is allowed to escape through the **partially reflecting mirror** to create the **laser beam**.

A **beam of laser light** having a **power** even as low as 0.1 mW may cause **eye damage**.

The diameter of a circular laser beam may stay at approximately 1 mm over a long distance. This means that it has a much higher irradiance than conventional light sources.

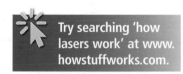
Try searching 'how lasers work' at www. howstuffworks.com.

Irradiance

School laser, power = 0.5 mW, beam diameter = 1 mm	Light bulb, power = 60 W, distance = 1 m (giving an approximate spherical area of 12 m²)
$I = \dfrac{P}{A} = \dfrac{P}{\pi r^2} = \dfrac{0.5 \times 10^{-3}}{3.14 \times (0.5 \times 10^{-3})^2}$ $= 636 \; Wm^{-2}$	$I = \dfrac{P}{A} = \dfrac{60}{12} = 5 \; Wm^{-2}$

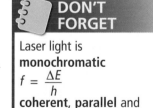
DON'T FORGET

Laser light is **monochromatic** $f = \dfrac{\Delta E}{h}$ **coherent**, **parallel** and has **high irradiance**.

LET'S THINK ABOUT THIS

1 The power from a point source spreads over a spherical area.

$I = \dfrac{P}{A} = \dfrac{P}{4\pi r^2} \Rightarrow I \propto \dfrac{1}{r^2}$

point source

2 The pupil of your eye has a larger area than the diameter of a laser beam!

sphere

SEMICONDUCTORS AND THE JUNCTION DIODE

SEMICONDUCTORS

Conductors, insulators and semiconductors

Conductors: Metals and semi-metals (graphite, antimony and arsenic) have **many free electrons**.

Insulators: In materials such as plastic, rubber, glass and wood, the electrons are all bonded and there are **no or few free electrons**.

Semiconductors: Silicon, germanium, selenium and some compounds have a **resistance** between good conductors and good insulators.

Pure semiconductors (intrinsic semiconductors) are **insulators** at very low temperatures. The addition of **heat**, **light** or a **voltage** causes a few **electrons** to escape from their atoms, leaving behind 'holes'. These can allow a small current to exist so that the resistance is decreased.

Pure silicon or **germanium** has four valence electrons in the outer shell, which **bond** with adjacent atoms so that there are no spare charge carriers.

B	C	N	O
Boron	Carbon	Nitrogen	Oxygen
Al	Si	P	S
Aluminium	Silicon	Phosphorus	Sulphur
Zn	Ga	Ge	As	Se
Zinc	Gallium	Germanium	Arsenic	Selenium
Cd	In	Sn	Sb	Te
Cadmium	Indium	Tin	Antimony	Tellurium
Hg	Tl	Pb	Bi	Po
Mercury	Thallium	Lead	Bismuth	Polonium

conductors
semiconductors
insulators

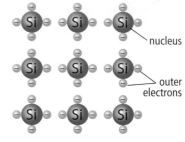

silicon as an insulator

Impurities for doping

Doping a semiconductor (extrinsic semiconductor) is the **addition** of a small amount of **impurity atoms** to a pure semiconductor, to **increase** the **electrical conductivity** of the semiconductor, **decreasing** its **resistance**.

N-type semiconductor

When a pure semiconductor with four valence electrons is **doped** with an **impurity** with **five** valence electrons, four form bonds and one electron is left as a **free charge carrier**.

The majority of the **free charge carriers** are **negative**, and this is now known as an **n-type** material. As the pure semiconductor and the impurity were both electrically neutral, the n-type material is also **electrically neutral**.

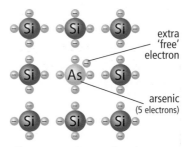

silicon as an n-type semiconductor

> **DON'T FORGET**
>
> Both **n-type** and **p-type** materials are **electrically neutral**.

P-type semiconductor

When the pure semiconductor with four valence electrons is **doped** with an **impurity** with **three** valence electrons, three bonds are formed and there is a 'hole', or missing electron.

This hole can move through the lattice structure, effectively carrying positive charge.

This is now known as a **p-type** material with a majority of **free holes** or **positive charge carriers**.

A hole moves when an electron jumps:

Before: e   e   e → o   e   e

After: e   e   o   e   e   e   What has moved?

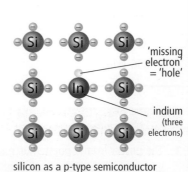

silicon as a p-type semiconductor

> **DON'T FORGET**
>
> N- and p-type materials are used to make **semiconductor devices** such as **diodes**, **transistors**, **LEDs** and **LDRs**.

> **DON'T FORGET**
>
> The number of impurity atoms used is very small.

THE JUNCTION DIODE

The p–n junction

A **diode** consists of a piece of **p-type** semiconductor in contact with an **n-type** semiconductor.

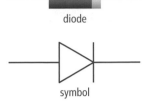

diode

symbol

Electrons drift from n-type to p-type to **fill adjacent holes**. This creates a **depletion layer** at the junction with **no free charge carriers**. (The **p** side of the junction gets a small negative charge and the **n** side of the junction gets a small positive charge creating a small voltage barrier ΔV against the further drift of charge.)

The depletion layer is an **insulator**.

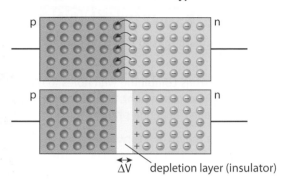

depletion layer (insulator)

> **DON'T FORGET**
>
> An **external voltage** applied to a p–n junction is called **bias**.

Forward and reverse bias

Forward bias	Reverse bias
<ul><li>**n**egative to **n**-type, **p**ositive to **p**-type</li><li>the diode **conducts**</li></ul>	<ul><li>**n**egative to **p**-type, **p**ositive to **n**-type</li><li>the diode **does not conduct**</li></ul>
The **depletion layer** is **reduced**. 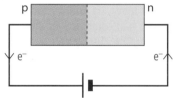	The **depletion layer widens**. 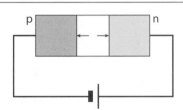
<ul><li>Electrons flow into the depletion layer from the n-type and then into the p-type.</li><li>Holes flow into the depletion layer from the p-type and then into the n-type.</li><li>Only a very small potential difference is needed (typically 0.5 V) to overcome the voltage barrier and the **diode conducts**.</li></ul>	<ul><li>Electrons in n-type are pulled by positive supply.</li><li>Holes in p-type are pulled by negative supply.</li><li>There is a wider area with no free charge carriers, a wider depletion layer and the diode **does not conduct**.</li></ul>
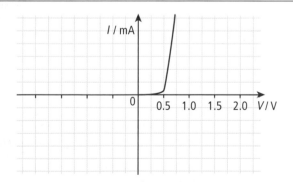	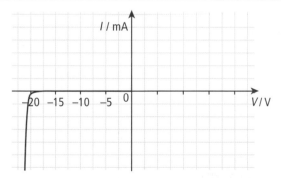
Forward bias: current when p.d. > 0.5 V (note the different voltage scales in graphs).	Reverse bias: minimal-leakage current (μA) until 'breakdown voltage' is reached.

LET'S THINK ABOUT THIS

A p–n diode is grown as a crystal with impurities grown in.

The depletion layer is about 10^{-6} m thick.

DIODES AND MOSFETS

THE LIGHT-EMITTING DIODE (LED)

An **LED** is a **forward-biased p–n junction diode**.

When the **LED conducts**, the emission of light is caused by **electrons combining** with **holes** to give out **energy** as **photons** of **light**. Each **electron–hole** recombination gives **one photon** out.

In the **junction region** of a **forward-biased** p–n junction diode, **positive** and **negative charge carriers** may **recombine** to give **quanta** of **radiation**.

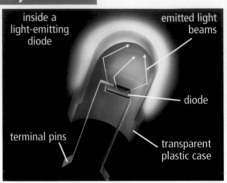

inside a light-emitting diode

emitted light beams

diode

terminal pins

transparent plastic case

THE PHOTODIODE

A **photodiode** is a solid-state device in which **positive** and **negative charges** are produced by the action of **light** on a **p–n junction**.

When a **photon** of light enters the **depletion layer**, its **energy** can be **absorbed** and an **electron–hole pair created**. Electrons, which had combined with holes to create the depletion layer, can now be **released** by photons of light. The number of **electron–hole** pairs **varies with** the number of **photons**. The photodiode can be used in two different modes.

Photovoltaic mode

In the **photovoltaic mode**, there is **zero** bias voltage applied. In fact, the photodiode acts as a **solar cell**.

In **photovoltaic mode**, a photodiode may be used to **supply power** to a load.

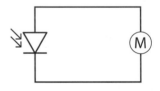

Photons which are incident on the depletion layer have their **energy absorbed**, freeing electrons and **creating electron–hole pairs**. These create a **voltage** or **e.m.f.** More **light** creates more **electron–hole** pairs.

- The voltage or e.m.f. is directly proportional to the light irradiance.
- Changes in irradiance produce rapid changes in voltage.

e.m.f. ∝ irradiance

Photoconductive mode

The **photodiode** is connected in **reverse bias** in photoconductive mode. The photodiode operates as an **LDR** or a **light sensor**. The photodiode will **not conduct** in the dark.

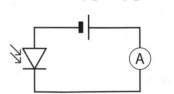

Photons of light shining into the depletion layer create **electron–hole pairs**. The free charge carriers released **lower the resistance** and **create** a small (leakage) **current**.

The greater the **irradiance of light**, the greater the **number of photons**, the more **free charge carriers** and the greater the **current**.

- The **current** does **not** depend on the supply voltage.
- The **current** is **directly proportional** to the light **irradiance**.
- Changes in irradiance produce **rapid changes** in current.

current ∝ irradiance

MOSFETS

A **MOSFET** is a **M**etal **O**xide **S**emiconductor **F**ield **E**ffect **T**ransistor.

The parts of a MOSFET

the substrate	p-type silicon semiconductor
n-region implants	grown by diffusion
silicon dioxide layer	insulating oxide layer is grown, etched to expose the n-type implants
metal contacts	made to source, gate, drain and substrate
s = source	n-type semiconductor (with connection to p-type substrate)
g = gate	insulated from substrate by oxide layer
d = drain	n-type semiconductor

n-channel MOSFET OFF

A MOSFET is connected in a circuit with the **drain** more **positive** than the source.

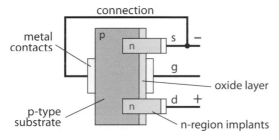

- **No voltage** applied to the **gate** (or $V_{GS} < 2V$).
- The junction between the **drain** and the **substrate** is **reverse**-biased (source – substrate unbiased).
- The **reverse bias** and the **high resistance** of the **substrate** prevent the flow of charge from s to d.

The MOSFET is OFF.

Transistor switch

If $V_{GS} > 2V$, I_D is **ON**, and the MOSFET conducts.

Below the threshold voltage, the MOSFET is **OFF**.

Amplifier

When the MOSFET is **ON**, the **gate voltage** V_{GS} controls the source-drain circuit **current** I_D.

As the gate voltage is increased, **additional electrons** help form a **wider n-channel** between the source and the drain.

The **drain current**, I_D, can be increased by altering V_{GS} or V_S.

The MOSFET can also **amplify voltage**: $A = \dfrac{V_0}{V_1} = \dfrac{V_{DS}}{V_{GS}}$

n-channel MOSFET ON

A positive **voltage** is now applied to the **gate**.

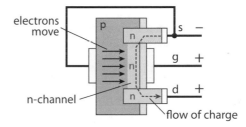

- The voltage $V_{GS} > 2V$ (**threshold voltage**).
- An **electric field** is set up between the gate and the substrate electrode, which draws up **electrons**.
- A **narrow channel** of **electrons** is formed between the source and the drain, allowing **charge to flow**.

As there is a **current**, the MOSFET is **ON**.

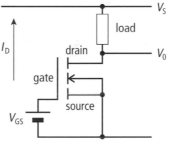

DON'T FORGET

A **transistor** can be an **electronic switch** or an **amplifier**.

LET'S THINK ABOUT THIS

1 An LED is built from a compound such as Gallium Arsenide Phosphide.

2 Red, orange, yellow, green, blue, infra-red and ultra-violet LEDs have been made.

3 The photodiode, in photovoltaic mode, is the reverse idea to the LED.

4 In the p-channel MOSFET, positive charge carriers are formed. The arrow on the symbol is reversed. The Higher course concentrates on the **n-channel enhancement MOSFET**.

ATOMIC MODELS AND REACTIONS

ATOMIC MODELS

The **atomic model** of the atom has been developed as a progression through many models, by famous physicists, to the model we know today.

400 BC.: The first person to believe in the existence of a smallest possible piece of matter was the ancient Greek philosopher **Democritus**. He called the particles 'atomos', meaning indivisible.

1803: **John Dalton** proposed the **Solid Sphere Model**: that elements are made of **atoms**.

1897: **J. J. Thomson**, on discovering 'corpuscles' (**electrons**), said: 'I can see no escape from the conclusion that cathode rays are charges of negative electricity carried by particles of matter'. He calculated their **charge-to-mass ratio** and found that electrons were **smaller** than the size of an atom.

1910: **J. J. Thomson** proposed the **plum-pudding model**, with thousands of tiny negatively charged corpuscles inside a cloud of massless positive charge.

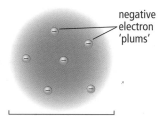
negative electron 'plums'

positive 'pudding'

1911: Thomson's own former student, **Ernest Rutherford**, came up with an improved theory, after having supervised two young colleagues (Geiger and Marsden) performing an experiment.

Geiger and Marsden's experiment

A stream of **alpha particles** (charge +2, mass 4amu) was fired at a **thin sheet** of **gold foil** in a vacuum.

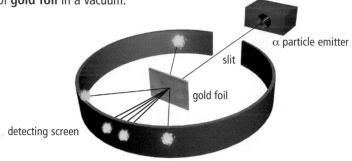
α particle emitter
slit
gold foil
detecting screen

Type *Geiger Marsden* or *Rutherford Model* into a search engine for more information.

Based on J. J. Thomson's model, many scattering angles were expected, but a different result was found:

- **most** particles **passed straight through**
- only a **few** were **deflected** through large angles.
- about 1 in 8000 **came backwards**!

Rutherford suggested the explanations:

- Most particles pass through the foil (>100 atoms thick), **so most of the atom must be empty space**.
- A few were deflected or bounced back, **so most of the mass and positive charge of an atom is concentrated in a very small volume**.

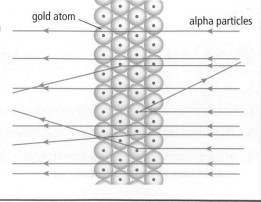
gold atom
alpha particles

Rutherford described it as the most incredible event of his life, 'as if you fired a 15-inch shell at a piece of tissue paper and it came back and hit you'.

DON'T FORGET

The history should help your understanding of physics, so learn Geiger and Marsden's experiment and Rutherford's model of the atom.

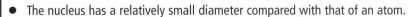

ATOMIC MODELS

1911: Planetary model or **nuclear model** proposed by Ernest Rutherford. **Rutherford** suggested that the **atom** might resemble a tiny solar system, with a **massive, positively charged centre** or **nucleus** circled by only a **few electrons**.

- The nucleus has a relatively small diameter compared with that of an atom.

- Most of the mass of an atom is concentrated in the nucleus.

1913: **Bohr model** or orbit model proposed by Niels Bohr. Based on the work you have studied with emission lines, photons and energy levels, Bohr confined **electrons** to orbit in **certain energy levels** or **shells**. The electrons **radiate energy** only in **transitions**.

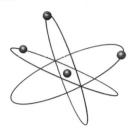

1924–6: **Electron cloud** or **quantum mechanical model** proposed by Louis de Broglie and Erwin Schrödinger. Electrons exist in wavelike orbitals. (This is beyond the Higher Physics course.)

nucleus

DECAY OF RADIONUCLIDES

The structure of the atom is based on **protons**, **neutrons** and **electrons**.

Particle	Mass (amu)	Charge	Symbol
proton	1	+1	p
neutron	1	0	n
electron	$\frac{1}{1840}$	−1	e

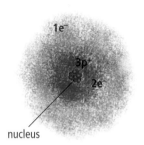

DON'T FORGET

mass number,
A = number of protons + neutrons.

$$_{26}^{56}\text{Fe}$$

atomic number,
Z = number of protons

A **radionuclide** is an isotope that **decays radioactively**.

Alpha, beta and **gamma radiations** all **emit** from the **nucleus** of an atom.

Radiation	Nature	Mass	Speed	Charge	Symbol	Deflection in a magnetic or electric field
alpha	He nucleus	4	slow	+2	α	yes
beta	fast electron	$\frac{1}{1840}$	0.9 c	−1	β	yes
gamma	e-m wave	0	c	0	γ	no

DON'T FORGET

Isotopes of an element have the **same atomic number** but **different mass number**.

Alpha decay: α 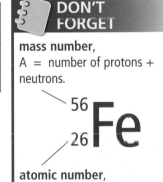 The daughter is a different element.

$$_Z^A X \rightarrow {}_{Z-2}^{A-4} Y + {}_2^4 \alpha$$

Beta decay: β A neutron decays into an electron (beta) and a proton.

$$_0^1 n \rightarrow {}_1^1 p + {}_{-1}^0 \beta$$

Gamma decay: γ No change in isotope, energy loss by e-m radiation.

 http://www.aip.org/history/electron/

LET'S THINK ABOUT THIS

1 Rutherford discovered the proton in 1918 and Chadwick the neutron in 1932.

2 Modern measurements show the average nucleus diameter about 10^{-15} m, 100 000 times smaller than the atom diameter.

FISSION AND FUSION

NUCLEAR FISSION

Fission is the **splitting** of a **large nucleus**.

In **fission**, a **nucleus** with a **large mass number splits** into **two nuclei** of **smaller mass numbers**, usually with the release of **neutrons**. **Energy** is **released**.

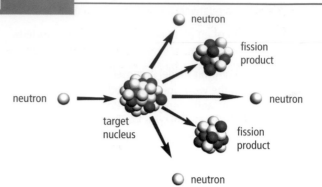

Spontaneous fission

Fission may be **spontaneous**, with each fission occurring at **random**, but the **half-life** will be **constant** for a large number of atoms.

$$^{256}_{100}Fm \rightarrow {}^{140}_{54}Xe + {}^{112}_{46}Pd + 4{}^{1}_{0}n + energy$$

Induced fission

Fission may be **induced** by **neutron** bombardment. An **incident neutron** can **stimulate** the **fission** of a **nucleus** with a **large mass number**.

In the following reaction, the U^{235} momentarily becomes U^{236}, but this is unstable and immediately undergoes fission.

$$^{1}_{0}n + {}^{235}_{92}U \rightarrow {}^{141}_{56}Ba + {}^{92}_{36}Kr + 3{}^{1}_{0}n + energy$$

Induced fission is used in the **reactors** in **nuclear power** stations.

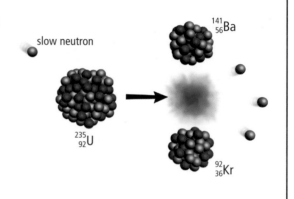

NUCLEAR FUSION

Fusion is the **joining** of **nuclei**.

In **fusion**, two nuclei **combine** to form a **nucleus** of **larger mass number**. The nuclei that fuse together are usually very **small**.

A large amount of **energy** is released, and **no radioactive waste** is produced in the reaction.

There is a **virtually unlimited** amount of the isotopes of hydrogen needed for **fusion** in seawater, and no greenhouse gases are emitted.

$$^{2}_{1}H + {}^{3}_{1}H \rightarrow {}^{4}_{2}He + {}^{1}_{0}n + energy$$

Fusion is the energy source of the **sun** and the **stars**, but physicists are still working on the design of **nuclear-fusion reactors** for Earth!

DON'T FORGET
Fission = splitting, fusion = joining

EINSTEIN AND ENERGY

During **fission** or **fusion**, the number of **nucleons** (particles of the nuclei) is the **same** before and after the reaction. The total **mass number** is conserved.

The **total mass** of the individual particles before and after the reaction is **not the same**. The **mass of the products** is always **less** than the **mass of the starting materials**. The missing mass is called the **lost mass**.

Einstein proposed that the **mass** and **energy** are **equivalent** and that the lost mass is turned into released energy, using his famous relationship:

$$E = mc^2$$

The **products** of **fission** and **fusion** acquire large amounts of **kinetic energy**.

Calculating the energy released by fission or fusion

- Find the difference in mass between each side of the equation.

- Calculate the energy released from the lost mass using Einstein's equation.

Example

How much energy is released from a single nucleus in the following fission reaction?

$$\mathrm{^{1}_{0}n} + \mathrm{^{235}_{92}U} \rightarrow \mathrm{^{138}_{55}Cs} + \mathrm{^{96}_{37}Rb} + 2\,\mathrm{^{1}_{0}n}$$

Mass before: $\mathrm{^{1}_{0}n}$ $1.6750 \times 10^{-27}\,\mathrm{kg}$
  $\mathrm{^{235}_{92}U}$ $3.9014 \times 10^{-25}\,\mathrm{kg}$ Total $= 3.91815 \times 10^{-25}\,\mathrm{kg}$

Mass after: $\mathrm{^{138}_{55}Cs}$ $2.2895 \times 10^{-25}\,\mathrm{kg}$
  $\mathrm{^{96}_{37}Rb}$ $1.5925 \times 10^{-25}\,\mathrm{kg}$
  $\mathrm{^{1}_{0}N}$ $1.6750 \times 10^{-27}\,\mathrm{kg}$
  $\mathrm{^{1}_{0}N}$ $1.6750 \times 10^{-27}\,\mathrm{kg}$ Total $= 3.9155 \times 10^{-25}\,\mathrm{kg}$

Lost mass $= 2.65 \times 10^{-28}\,\mathrm{kg}$

Energy released: $E = mc^2 = 2.65 \times 10^{-28}(3.0 \times 10^{8})^2 = 2.38 \times 10^{-11}\,\mathrm{J}$

Example

There are $\dfrac{1}{3.9014 \times 10^{-25}} = 2.56 \times 10^{24}$ atoms in 1 kg of uranium.

The energy in 1 kg is therefore $(2.385 \times 10^{-11})(2.56 \times 10^{24}) = 6.11 \times 10^{13}\,\mathrm{J}$

1 kg of coal releases approximately $3 \times 10^7\,\mathrm{J}$. That is about 2 million times less than 1 kg of uranium!

DON'T FORGET

Take care not to drop figures in lost-mass calculations – they are all significant!

DON'T FORGET

Nuclear-fusion calculations are done in the same way.

  Try entering *nuclear fusion* or *nuclear fission* into an internet search engine.

LET'S THINK ABOUT THIS

1   In **nuclear fission**, the three **neutrons** released travel **too fast** to cause **further fission**, so in a **nuclear reactor** they are **slowed** first by travelling through a **moderator**. **Control rods** are also used to **absorb neutrons** to control a **chain reaction**.

2   The sun's **fusion** takes place at extremely **high temperatures**. The job for the next generation of physicists (that's you!) is to make **cold fusion** happen. This would solve a lot of the **energy problems** of the world.

MEASURING RADIATION

ACTIVITY

Radioactive decay is a **random** process. A radioactive substance contains **many nuclei**, which decay at random. It is impossible to predict when an individual nucleus will decay, but with so many nuclei in even a small sample we can predict the average **number** that will decay in a certain **time**.

The **rate of decay** is known as the **activity** of the substance:

$$activity = \frac{number\ of\ nuclei\ decaying}{time} \qquad A = \frac{N}{t}$$

The **activity** of a radioactive source is the **number of decays** per **second**.

Radioactivity is measured in **becquerels (Bq)**. One becquerel is one decay per second.

Example

The activity of 1 g of uranium-238 is 12 kBq. How many decays will occur in 1 minute?

Activity $= N \times t = 12\,000 \times 60 = 720\,000$ decays.

Half-life

Once decay has occurred, there are fewer nuclei available to decay.

The **time taken** for the **activity** of a particular sample to **halve** is called the **half-life**.

The half-life of this substance is 0.7 hours.

Every 0.7 hours, the number of radioactive nuclei drops by half.

Fractional activity
$= 1, \frac{1}{2}, \frac{1}{4}, \frac{1}{8}, \cdots$

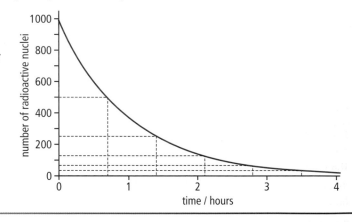

ABSORBED DOSE

When **ionising radiation** is absorbed by the human body, scientists consider the **energy of the absorbed particles** and the **mass of matter** absorbing the radiation.

Absorbed dose is the **energy** absorbed **per unit mass** of the absorbing material.

$$absorbed\ dose = \frac{energy}{mass} \qquad D = \frac{E}{m}$$

The gray, **Gy**, is the **unit** of absorbed dose. One gray is one joule per kilogram. $1\ Gy = 1\ Jkg^{-1}$

Example

If a man had his whole body irradiated with a gamma radiation until he had received 10 J energy, calculate his absorbed dose.

$$D = \frac{E}{m} = \frac{10}{90} = 0.11\ Jkg^{-1} = 0.11\ Gy$$

For comparison, a chest X-ray might deliver 0.3 mGy.

DON'T FORGET

Activity is from the source; count rate is what is detected.

RADIATION WEIGHTING FACTOR

The absorbed dose is only measuring the energy deposited in the tissue and does not take into account the effects of different types of radiation.

A **radiation weighting factor, w_R,** is given to each kind of radiation as a measure of its biological effect:

X and γ rays	1
β particles	1
α particles	10
heavier nuclei	20
protons	10
slow neutrons	3
fast neutrons	10

EQUIVALENT DOSE

Equivalent dose makes allowance for the type of radiation.

The **equivalent dose, H,** is the **product** of **absorbed dose** and **radiation weighting factor**.

Equivalent dose is measured in **sieverts (Sv)**. $H = Dw_R$

The same **equivalent dose** always gives the same **biological effect**. For example, we now know that 1 mSv of γ radiation will do the same damage as 1 mSv of α radiation.

Equivalent dose rate

We need to monitor the rate at which radiation is absorbed by people who work with radiation over periods of time.

The **equivalent dose rate** is the **equivalent dose** per **unit time**.

$$\dot{H} = \frac{H}{t}$$

The units of **equivalent dose rate** can be $mSv\,h^{-1}$ or $mSv\,yr^{-1}$ depending on the units of time.

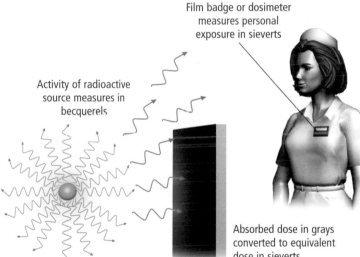

Activity of radioactive source measures in becquerels

Film badge or dosimeter measures personal exposure in sieverts

Absorbed dose in grays converted to equivalent dose in sieverts

Example

If the man in the example on p. 88 received his absorbed dose over 5 minutes, his equivalent dose rate would be:

$$H = Dw_R = 0.11 \times 1 = 0.11\,Sv$$

$$\dot{H} = \frac{H}{t} = \frac{0.11}{5 \times 60} = 0.37\,mSv\,s^{-1}$$

DON'T FORGET

Activity in Bq.
Absorbed dose in Gy.
Equivalent dose in Sv.

LET'S THINK ABOUT THIS

1 Alpha radiation is absorbed by a thin layer of tissue, which gives a **high** absorbed dose.

2 Depending on which organ of the body receives radiation, the medical physicist will multiply the equivalent dose by **another weighting factor** to take into account the susceptibility of **different organs or tissues** to damage from the radiation. This value is still measured in Sv but is called the **effective equivalent dose**.

PROTECTION

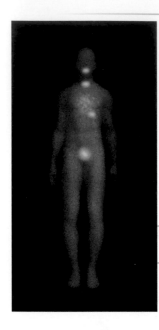

BIOLOGICAL HARM

Ionising radiation will cause **harm** to living cells and can **destroy** these cells.

Alpha particles do not penetrate the body from outside but they are harmful if taken internally. **Beta particles** can penetrate about 1 cm of tissue and can be very harmful. **X-rays** and **gamma-rays** produce little ionisation but can pass through the body from outside or inside and can damage organs. They can be useful for diagnostic scans and in shrinking tumours.

The **risk of biological harm** from an exposure to radiation depends on:

- the **absorbed dose** (which depends on the energy of radiation and mass of tissue)
- the **kind of radiations** (e.g. X- or γ-rays, β- or α-particles, slow neutrons etc.)
- the **body organs** or **tissues** exposed.

BACKGROUND RADIATION

Background radiation is all around us and has to be deducted from any measurements or radiation sources.

Corrected count s^{-1} = measured count s^{-1} − background count s^{-1}

Background radiation comes from two types of source: **natural** and **artificial**.

Natural source	Annual equivalent dose	
	μSv	mSv
Radioactive gases in air and buildings (radon and thoron)	800	0.80
Rocks of the Earth	400	0.40
In food and our bodies	370	0.37
Cosmic rays from space	300	0.30
Total natural sources	1870	1.87

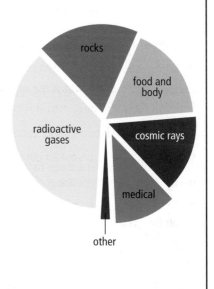

Artificial source	Annual equivalent dose	
	μSv	mSv
Medical uses (X-rays)	250	0.250
Weapons testing	10	0.010
Nuclear industry (waste)	2	0.002
Other (job, TV, flights)	18	0.018
Total man-made sources	280	0.280

The **average annual effective dose** that a person in the UK receives due to natural sources (cosmic, terrestrial and internal radiation) is approximately **2 mSv**.

Annual effective dose limits

Annual effective dose limits have been set for exposure to radiation for the **general public**, and **higher limits** for workers in **certain occupations**.

The annual limits are:

1 mSv yr^{-1} for the public and 20 mSv yr^{-1} for radiation workers.

These limits are in addition to background radiation.

DON'T FORGET

The effective dose from background radiation is about **2 mSv yr^{-1}**.

RADIATION SAFETY

One of the simplest methods of reducing **equivalent dose rate** from ionising radiation is by **shielding**, or placing an **absorbing material** in the path of the radiation. Aluminium, lead, concrete and water have all been used to absorb radiation. A lead apron is worn by medical radiologists.

Increasing the **distance** from the source can **reduce equivalent dose rate** from ionising radiation. As radiation spreads out in all directions, it obeys the **inverse square law**, the same as light. This means that increasing distance has a greater effect than might be expected.

HALF-VALUE THICKNESS

The **half-value thickness** of an **absorber** for a particular radiation is the **thickness** that **absorbs half** of the radiation. Each time an additional half-value **thickness** is added, the count rate will **decrease by half**. Fractional count rate $= 1, \frac{1}{2}, \frac{1}{4}, \frac{1}{8}, \ldots$

Measuring the half-value thickness for an absorber:

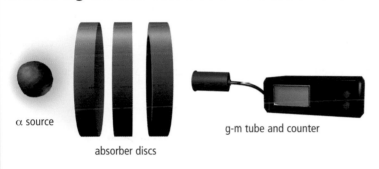

α source

absorber discs

g-m tube and counter

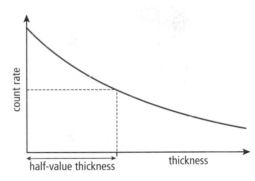

count rate

half-value thickness

thickness

- Without the source, measure the **background count rate**.

- Set the **source** a **fixed distance** in front of the detector and measure the count rate again.

- Place **plates** of the **material** being investigated **between** the source and the detector to make **different** measured **thicknesses**, and each time measure the count rate again.

- **Subtract** the **background count rate** from the readings to obtain the corrected count rates.

- A **graph** of the **corrected count rate** against **thickness** is drawn.

- The **thickness** that causes the **count rate** to **halve** is noted from the graph.

- To deal with **uncertainties**, the half-value thickness should be obtained from various places on the graph. The **average value** and **random uncertainty** can then be calculated (see pp. 6–9).

Example

Source 40 Sv h^{-1}. Half-value thickness is 8 mm. What is the equivalent dose rate after adding a 32 mm absorber? $40 \Rightarrow 20 \Rightarrow 10 \Rightarrow 5 \Rightarrow 2.5\,\text{Sv}\,\text{h}^{-1}$.

> **DON'T FORGET**
>
> You are expected to be able to describe the **principles** of a method for **measuring** the **half-value thickness** of an **absorber**.

 Type *radiation units* into an internet search engine for more information.

LET'S THINK ABOUT THIS

1 Radiation obeys the inverse square law for irradiance $I \propto \frac{1}{d^2}$, and similar equations for equivalent dose will be obeyed, i.e. $H \propto \frac{1}{d^2}$

2 Exams: Protection from the exam comes from being prepared. Well done for reaching this far and studying hard. Practise plenty of past exam questions in the run-up to the exam, and you are sure to do well.

KEY QUESTIONS

REVISION QUESTIONS

Waves

1 How does the frequency of a wave depend on its source?

2 What is the test for a wave?

3 Name at least four behaviours of waves.

4 If the path difference to the second order maxima in an interference pattern is 1000 nm, what is the wavelength of the radiation?

5 State approximate values for the wavelengths of red, green and blue light.

6 What does the energy of a wave depend on?

7 What is the meaning of the term in phase?

8 What is the meaning of the term coherent?

9 Describe the principles of a method for measuring the wavelength of a monochromatic light source, using a grating.

Refraction of light

10 State Snell's law.

11 A ray of light enters a glass block at 42° from the normal. If the angle of refraction in the glass is 22°, what is the refractive index of this glass?

12 What velocity will the ray in question 11 travel at in the glass?

13 Explain what is meant by critical angle θ_c.

14 Why does a medium have a range of refractive indexes?

15 Explain what is meant by total internal reflection.

16 Derive the relationship between critical angle and absolute refractive index of a medium.

Optoelectronics and semiconductors

17 Explain what is meant by irradiance.

18 What conditions are required for photoelectric emission to occur?

19 State the relationship between photoelectric current and irradiance.

20 Describe the nature of a beam of radiation.

21 Calculate the energy of a photon of wavelength 500 nm.

22 How do you calculate the maximum kinetic energy of photoelectrons?

23 Where do you find electrons in a free atom?

24 Explain the occurrence of absorption lines in the spectrum of sunlight.

25 What is spontaneous emission of radiation?

26 Explain the function of the mirrors in a laser.

27 Give examples of conductors, insulators and semiconductors.

28 How is radiation emitted from a p–n junction diode?

29 What is a photodiode?

30 When does a photodiode leakage current occur?

31 What can an n-channel enhancement MOSFET be used for?

Nuclear reactions

32 Describe Rutherford's model and how he derived it.

33 State what is meant by alpha, beta and gamma decay of radionuclides.

34 Explain fusion.

35 Identify the processes occurring in nuclear reactions written in symbolic form.

36 Explain fission.

Dosimetry and safety

37 Define activity and its unit.

38 What is a radiation weighting factor for?

39 Define equivalent dose and give its unit. What is equivalent dose rate?

contd

REVISION QUESTIONS contd

40 What does the risk of biological harm from an exposure to radiation depend on?

41 Describe the factors affecting the background radiation level.

42 What is the average annual effective dose that a person in the UK receives?

43 How does the intensity of a beam of gamma radiation vary with the thickness of an absorber.

ANSWERS

1 The frequency of a wave is the same as the frequency of the source producing it. **2** Interference.

3 Reflection, refraction, diffraction and interference are characteristic behaviours of all types of waves.

4 500 nm. **5** red 644 nm   green 509 nm   blue 480 nm

6 The energy of a wave depends on its amplitude. **7** The amplitudes are in step.

8 Coherent waves have the same frequency, wavelength and speed along with a constant phase relationship. **9** See p 67–68.

10 The ratio $\sin\theta_1 / \sin\theta_2$ is a constant when light passes obliquely from medium 1 to medium 2. The absolute refractive index, n, of a medium is the ratio $\sin\theta_1 / \sin\theta_2$ where θ_1 is in a vacuum (or air as an approximation) and θ_2 is in the medium.

11 1.79. **12** $1.68 \times 10^8\,\mathrm{ms^{-1}}$. **13** See p 72.

14 The refractive index depends on the frequency of the incident light.

15 See p 72. **16** See p 72.

17 The irradiance at a surface on which radiation is incident is the power per unit area.

18 Photoelectric emission from a surface occurs only if the frequency of the incident radiation is greater than some threshold frequency f_0, which depends on the nature of the surface. For frequencies smaller than the threshold value, an increase in the irradiance of the radiation at the surface will not cause photoelectric emission.

19 State that for frequencies greater than the threshold value, the photoelectric current produced by monochromatic radiation is directly proportional to the irradiance of the radiation at the surface.

20 A beam of radiation can be regarded as a stream of individual energy bundles called photons, each having an energy dependent on the frequency of the radiation.

21 4×10^{-19} J.

22 Photoelectrons are ejected with a maximum kinetic energy which is given by the difference between the energy of the incident photon and the work function of the surface.

23 Electrons in a free atom occupy discrete energy levels. **24** See p 77.

25 Spontaneous emission of radiation is a random process analogous to the radioactive decay of nucleus. **26** See p 79. **27** See p 80.

28 In the junction region of a forward-biased p–n junction diode, positive and negative charge carriers may recombine to give quanta of radiation.

29 A photodiode is a solid-state device in which positive and negative charges are produced by the action of light on a p–n junction.

30 The leakage current of a reverse-biased photodiode is directly proportional to the light irradiance and fairly independent of the reverse-biasing voltage, below the breakdown voltage.

31 An n-channel enhancement MOSFET can be used as an amplifier.

32 Rutherford showed that: a) the nucleus has a relatively small diameter compared with that of the atom; b) most of the mass of the atom is concentrated in the nucleus. See also p 84.

33 See p 85. **34** In fusion two nuclei combine to form a nucleus of larger mass number. **35** See p 86.

36 In fission a nucleus of large mass number splits into two nuclei of smaller mass numbers, usually with the release of neutrons. Fission may be spontaneous or induced by neutron bombardment.

37 The activity of a radioactive source is the number of decays per second and is measured in becquerels (Bq), where one becquerel is one decay per second.

38 A radiation weighting factor is given to each kind of radiation as a measure of its biological effect.

39 The equivalent dose is the product of absorbed dose and radiation weighting factor and is measured in sieverts (Sv). Equivalent dose rate is the equivalent dose per unit time.

40 The risk of biological harm from an exposure to radiation depends on: **a)** the absorbed dose. **b)** the kind of radiations, eg α, β, γ, slow neutron. **c)** the body organs or tissues exposed.

41 See p 90.

42 The average annual effective dose that a person in the UK receives due to natural sources (cosmic, terrestrial and internal radiation) is approximately 2 mSv. Annual effective dose limits have been set for exposure to radiation for the general public, and higher limits for workers in certain occupations.

43 Sketch a graph as at p 91.